How to Rebuild GM LS-SERIES ENGINES

Chris Werner

CarTech®

CarTech®

CarTech,® Inc.
6118 Main Street
North Branch, MN 55056
Phone: 651-277-1200 or 800-551-4754
Fax: 651-277-1203
www.cartechbooks.com

Edit by Rick Jensen
Layout by Monica Bahr

ISBN 978-1-932494-60-0
Item No. SA147

Library of Congress Cataloging-in-Publication Data

Werner, Chris
How to rebuild GM LS-series engines / by Chris Werner.
p. cm. -- (Workbench series) (S-A design)
Includes bibliographical references.
ISBN 978-1-932494-60-0
1. General Motors automobiles--Motors--Maintenance and repair. 2. General Motors automobiles--Performance. 3. General Motors automobiles--Motors--Modification. I. Title. II. Series.

TL215.G43W38 2009
629.25'04--dc22

2009011578

Written, edited, and designed in the U.S.A.
Printed in China
20 19 18 17 16

Back Cover Photos

Top Left: *This engine was installed in a 2004 Corvette Z06, and while the wiring harness and many accessories have already been taken off as part of the process of its removal, other accessories are still intact.*

Top Right: *The camshaft on any pushrod V-8 is bulky and long, without anything to grab onto. To remedy this problem we have inserted a 3/8 extension into the hollow bore in the center of the camshaft.*

Middle Left: *The Gen III LQ9 used iron blocks for enhanced durability. It is a great candidate to reuse on a new project vehicle; don't let a little outside surface rust turn you off.*

Middle Right: *These counterweights on an aftermarket LS crank show two holes that have been drilled perpendicular to the crankshaft's centerline (material removed), as well as one metal slug that has been inserted (material added).*

Bottom Left: *Inspect all areas of the block for any contamination. If you find much debris, take the block back to the machine shop for a jet wash.*

Bottom Right: *Install the main caps (with dry bearings) and tighten them in the proper sequence. One good tip: Do not use a hammer to align the thrust bearing surfaces.*

DISTRIBUTION BY:

Europe
PGUK
63 Hatton Garden
London EC1N 8LE, England
Phone: 020 7061 1980 • Fax: 020 7242 3725
www.pguk.co.uk

Australia
Renniks Publications Ltd.
3/37-39 Green Street
Banksmeadow, NSW 2109, Australia
Phone: 2 9695 7055 • Fax: 2 9695 7355
www.renniks.com

Canada
Login Canada
300 Saulteaux Crescent
Winnipeg, MB, R3J-3T2 Canada
Phone: 800 665 1148 • Fax: 800 665 0103
www.lb.ca

TABLE OF CONTENTS

DEDICATION

This book is dedicated to the memory of Alan F. Werner, 1920–2001.

ACKNOWLEDGMENTS

Without the support and generosity of the following people, this book could not have been possible:

George Benson; Lou & Shawna Bengivenni, LRB Performance Machine Company; The staff of CarTech; The staff of Bill Ceralli Competition Engines; Hank Daniecki, SLP Performance Parts; Rick Jensen, Editor, *GM High-Tech Performance*; Allison Keller; Cheryl McCarron, GM; Dr. Jamie Meyer, GM Performance Parts; Dave Monyhan, Goodson; Tom Read, GM Powertrain; Brian Reese, COMP Performance Group; Ron Rotunno, Federal-Mogul; Matt Sorian & Nick Stevko, TT Performance Parts; Fred Tobes, the Crank Shop; and the entire Werner family.

AUTHOR BIO

Since 1998, Chris Werner has been a regular contributor to several enthusiast automotive publications, most notably *GM High-Tech Performance*. Modifying small-block-powered GMs since his teenage years, his interests led him to seek a degree in Mechanical Engineering. After graduating with honors from Clemson University, he attended law school at the University of North Carolina at Chapel Hill. Now an attorney admitted to practice in New York, New Jersey, and before the United States Patent and Trademark Office, he specializes in automotive lemon, warranty, and product liability law. He still manages to spend every spare waking moment fueling his true passion, namely, upping the performance of GM muscle cars.

What is a *Workbench*® Book?

This Workbench® Series book is the only book of its kind on the market. No other book offers the same combination of detailed hands-on information and revealing color photographs to illustrate engine rebuilding. Rest assured, you have purchased an indispensable companion that will expertly guide you, one step at a time, through each important stage of the rebuilding process. This book is packed with real world techniques and practical tips for expertly performing rebuild procedures, not vague instructions or unnecessary processes. At-home mechanics or enthusiast builders strive for professional results, and the instruction in our Workbench® Series books help you realize pro-caliber results. Hundreds of photos guide you through the entire process from start to finish, with informative captions containing comprehensive instructions for every step of the process.

Appendixes located in the back of the book provide essential specification and rebuild information. These include diagrams and charts for cylinder firing order, torque sequences and specifications, piston ring gap alignment, and timing belt/chain alignment. In addition, general engine specifications, including compression ratio, bore and stroke, oil pressure, and many other specifications, are included.

The step-by-step photo procedures also contain many additional photos that show how to install high-performance components, modify stock components for special applications, or even call attention to assembly steps that are critical to proper operation or safety. These are labeled with unique icons. These symbols represent an idea, and photos marked with the icons contain important, specialized information.

Here are some of the icons found in Workbench® books:

Important!—Calls special attention to a step or procedure, so that the procedure is correctly performed. This prevents damage to a vehicle, system, or component.

Save Money—Illustrates a method or alternate method of performing a rebuild step that will save money but still give acceptable results.

Torque Fasteners—Illustrates a fastener that must be properly tightened with a torque wrench at this point in the rebuild. The torque specs are usually provided in the step.

Special Tool—Illustrates the use of a special tool that may be required or can make the job easier (caption with photo explains further).

Performance Tip—Indicates a procedure or modification that can improve performance. Step most often applies to high-performance or racing engines.

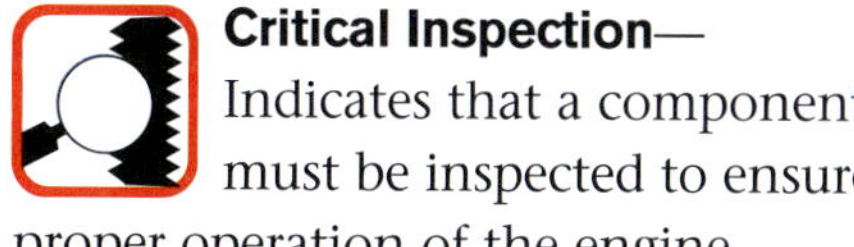

Critical Inspection—Indicates that a component must be inspected to ensure proper operation of the engine.

Precision Measurement—Illustrates a precision measurement or adjustment that is required at this point in the rebuild.

Professional Mechanic Tip—Illustrates a step in the rebuild that non-professionals may not know. It may illustrate a shortcut, or a trick to improve reliability, prevent component damage, etc.

Documentation Required—Illustrates a point in the rebuild where the reader should write down a particular measurement, size, part number, etc. for later reference or photograph a part, area or system of the vehicle for future reference.

Tech Tip—Tech Tips provide brief coverage of important subject matter that doesn't naturally fall into the text or step-by-step procedures of a chapter. Tech Tips contain valuable hints, important info, or outstanding products that professionals have discovered after years of work. These will add to your understanding of the process, and help you get the most power, economy, and reliability from your engine.

CHAPTER 1

The Small-Block Chevrolet Reborn: GM's Gen III and IV V-8s

(Illustration courtesy of General Motors)

So you're thinking about rebuilding your Gen III/IV General Motors small-block? Congratulations—an engine rebuild can be a very rewarding task. Right now, the prospect of enjoying a freshened engine probably excites you most; but we hope that with the help of this book, you'll enjoy the actual rebuilding process as well!

Before beginning the planning stages of your project, there are a few questions you'll need to ponder. Not the least of which is, what is a Gen III/IV? And how do you know that the engine you're looking to rebuild really is a Gen III or IV GM engine—or even a small-block at all? After all, the pioneering Gen III engine architecture has only been in existence since model year 1997—and its close cousin the Gen IV is even younger. While hundreds of thousands of Gen III and IV engines have been installed into production GM cars and trucks since the late 1990s, production year overlap with other GM V-8 engines means that while it's likely that your late-model's mill falls within the family, there is also a chance that it does not.

This chapter shows you what a Gen III/IV is and tells you a little about what makes it such an excellent engine design. We also provide you with information that will help you determine whether you've got one. But first, to put things in perspective, let us discuss how this engine architecture evolved from GM engines of years past.

The Small-Block Family Tree

In the decades leading to the 21st century, Detroit produced an incredible number of V-8 engines. While each manufacturer had experimented with different valvetrain configurations on its V-8s from time to time, by far the most common style was the so-called pushrod V-8. This type of engine, whose valves are actuated by a cam encased in the engine block, was originally known simply as an overhead valve (OHV) V-8. The success of this type of engine was such that it became a staple for most American-made automobiles.

Within General Motors, pushrod V-8s were made in a staggering variety. Numerous versions of Pontiac, Oldsmobile, Buick, Chevrolet, GMC, and Cadillac engines were installed in many cars and trucks, with a few engine types even overlapping into GM vehicles outside their respective division. Each engine design had its own particular advantages and quirks, but over time (and in the interest of cost cutting and cross-compatibility) only those V-8s GM considered the best of the best were to continue on in production. In the final decades of the 20th century, many V-8 engine lines were discontinued; and by the early 1990s, this left only small- and big-block Chevrolets as the pushrod gasoline V-8 engines used in all GM cars and trucks.

It was the small-block variant of the Chevrolet engine that would eventually form the inspiration for the Gen III. This, then, begs the question: "What is a small-block Chevy?"

Generation I

One of the most successful and versatile engines of all time, the original small-block debuted in 1955 as the Turbo-Fire—Chevrolet's first OHV V-8. Tasked with the development of a powerful upgrade for the "Stovebolt" six-cylinder, a Chevrolet engineering team (led by none other than the legendary Ed Cole) came up with this engine's compact, easy-to-manufacture design. Features like 4.4-inch bore spacing, a single-piece intake manifold, and an internal lubrication system contributed to a lightweight package, while wedge-shaped combustion chambers and low-mass, stamped-steel rocker arms helped yield a broad performance curve and high RPM potential. Powering everything from Corvettes to pickups, the small-block quickly earned a reputation for power and durability.

While in the years following it would spawn many engine displacements and undergo several design finesses (like changes in bearing sizes and a switch to a one-piece rear main seal), the small-block Chevrolet still kept to the same basic architecture as the original 265-ci version. The versatile small-block saw it all: from the muscle car wars of the late 1960s, to the suffering of the 1970s gas crisis, to the eventual use of electronic fuel injection beginning in the 1980s. The engine was so successful that it eventually was used almost without exception throughout all GM divisions, and since it represented the first generation of what would eventually be the corporate GM pushrod V-8, it is now best to refer to the beloved original small-block Chevy simply as the Gen I.

Confined to only light truck and van use in its last decade or so, the final production-vehicle-bound Gen I came off the assembly line in 2002. This marked an incredible run of nearly 50 years—making this engine one of the most successful of all time for any company. As a testament to its enduring legacy, it continues to be used in some marine applications and heavy-duty trucks—and high-performance, Gen I-based engines are still available to this day from GM Performance Parts.

Although technically not a change to the engine architecture itself, the Gen I small-block began receiving modern port fuel injection beginning in the mid-1980s. Shown here is a cutaway model equipped with the distinctive Tuned Port Injection (TPI) system first installed on 1985 Corvettes. Though only available in 305 and 350-ci versions in its later years, earlier carbureted Gen I engines had displaced as much as 400 ci. (Photo courtesy of General Motors)

The Gen I powered an amazing array of GM vehicles during its decades of use, making it one of the most respected engines of all time. However, you probably wouldn't ask this owner to see the "Gen I" under his Corvette's hood; the preferred term is still "small-block Chevy!"

Generation II

Despite several changes to the original small-block Chevrolet engine design over the years—some fairly substantial—GM never saw fit to come up with a new "generation" designation for its popular pushrod V-8. This all changed when the Gen II version was released in the early 1990s. This revised small-block is perhaps better known by the name of its most popular variant, the LT1.

Introduced in the 1992 Corvette, this short-lived engine family's most defining feature was its reverse-flow cooling system. Sending coolant to the cylinder heads before the engine block was said to enable a higher compression ratio and improved efficiency. Also, unlike the

First used in the 1955 model year, the original small-block Chevrolet displaced just 265 ci (4.3L)—quite small compared to later iterations. Though excellent for the time, its initial horsepower was relatively low, too—only 195 hp was mustered by the high-output Corvette version. Its designers could not possibly have imagined the huge power increases that the coming years would bring, nor the great legacy this engine design would create! (Photo courtesy of General Motors)

GM deemed it worthwhile to call its reverse-cooled small-block variant the Gen II, but the actual architecture of this engine family was not far removed from that of the original small-block Chevrolet. Certain distinctions like a front-mounted distributor and gear-driven water pump do set it apart, however. Pictured is the most common Gen II, the LT1. (Illustration courtesy of General Motors)

rear-mounted distributor of the original small-block, the Gen II used a front-mounted unit known as the Opti-Spark (a misnomer if there ever was one—this design has proven itself to be exceptionally unreliable). Other slight changes to the engine block, cylinder heads, and other components meant that very few engine parts were actually shared with the Gen I.

The LT1's name was a throwback to a high-power variant of the original small-block used in the early 1970s called the LT-1 (note the hyphenation difference). Sporting the same 350 ci as its namesake, the LT1's high-performance car use was confined to the 1992–1996 Chevrolet Corvette and V-8 versions of the 1993–1997 F-body (i.e., Chevrolet Camaro and Pontiac Firebird). Aside from the L99, the only other engine in the Gen II family was the LT4. Simply a higher-output version of the LT1, it was produced primarily in 1996 for manual-transmission-equipped Corvettes, with a few more being thrown into SLP-modified F-bodies available through dealerships in 1997.

Gen II small-blocks were never used in any GM trucks, and further indicative of this engine family's unique nature, few of the design changes it incorporated would be used subsequently in future small-block generations. It also fell out of production at a time when versions of the Gen I continued to be built. In short, the Gen II is a unique animal, and since it was not a ground-up redesign, it's perhaps best thought of as a special version of the original small-block and not as a completely separate engine. Yet, we'll call it a Gen II—if only because GM did.

The Gen III Era Begins

By the mid 1990s, 40 years of small-blocks had earned Chevrolet and GM a reputation for building some of the most versatile and durable V-8s ever made. But for all the variations within the first small-block generation (and even with a few Gen II motors thrown in), the same 1950s-era architecture had largely been retained. New corporate and federal guidelines for performance, manufacturability, durability, and emissions were looming on the horizon—it was time for a change.

A clean-slate design, the Gen III engine family would be the first truly new small-block since the original 1955 version, and its introduction marks the one true delineation between small-blocks of years past and those of today. The first Gen III engine was a 5.7L version introduced in the newly redesigned 1997 Corvette. Referred to as RPO LS1, this 346-ci wonder mill gained an enormous high-performance following almost immediately—leading enthusiasts to refer to all Gen III (and later, Gen IV) engines as simply "LS1s" or the "LS family." Though confined initially to use in sports cars, the new engine architecture quickly spread across the full gamut of GM cars, light trucks, and SUVs sold in the U.S., completely replacing older small-block V-8s within only a few years of its introduction. This is to say nothing of the extensive use of these engines in GM vehicles sold abroad (Australia's Holden nameplate and the U.K.'s Vauxhall brand are just two of many), proving that this engine family's appeal and

The 32-valve, 5.7L LT5 engine that powered the Corvette ZR-1 from 1990 to 1995 was also sometimes identified by the term "Gen II"—but with features like overhead cams, it had virtually nothing in common with the other engines known as Gen IIs—or, for that matter, any other generation of small-block.

Iron-headed versions of the LT1 were also used from 1994 to 1996 in some of GM's large cars, which included the Buick Roadmaster, Cadillac Fleetwood, and Chevrolet Caprice (including the high-performance version of the Caprice—the Impala SS—pictured here). The Caprice was also available with a small-bore, 4.3L version of the Gen II called the L99.

Gen III/IV-Equipped U.S. Passenger Cars

Common Name	Model Years	RPO	Displacement, Liters (CI)	Bore, mm (in)	Stroke, mm (in)	Block Material	Head Material	Compression Ratio	Horse-power**	Notes
GEN III:										
Vortec 4.8L	1999-2007	LR4	4.8 (293.4)	96.01 (3.780)	83.00 (3.268)	FE	AL	9.5:1	285	
Vortec 5.3L	1999-2007	LM7	5.3 (325.2)	96.01 (3.780)	92.00 (3.622)	FE	AL	9.5:1	295	
	2002-2007	L59	5.3 (325.2)	96.01 (3.780)	92.00 (3.622)	FE	AL	9.5:1	295	Flex Fuel
	2003-2004	LM4	5.3 (325.2)	96.01 (3.780)	92.00 (3.622)	AL	AL	9.5:1	300	
	2005-2007	L33	5.3 (325.2)	96.01 (3.780)	92.00 (3.622)	AL	AL	9.9:1	310	
LS1	1997-2004	LS1	5.7 (345.7)	99.00 (3.898)	92.00 (3.622)	AL	AL	10.1:1	350	The original Gen III
LS6	2001-2005	LS6	5.7 (345.7)	99.00 (3.898)	92.00 (3.622)	AL	AL	10.5:1	405	Higher-output version of the LS1
Vortec 6.0L	1999-2007	LQ4	6.0 (364.1)	101.6 (4.000)	92.00 (3.622)	FE	FE/AL	9.4:1	335	Pre-2001
versions had iron heads										
	2002-2007	LQ9	6.0 (364.1)	101.6 (4.000)	92.00 (3.622)	FE	AL	10.0:1	345	
GEN IV:										
Vortec 4.8L	2007-2009	LY2	4.8 (293.4)	96.01 (3.780)	83.00 (3.268)	FE	AL	9.1:1	295	
	2010-	L20	4.8 (293.4)	96.01 (3.780)	83.00 (3.268)	FE	AL	8.8:1	302	Flex Fuel
Vortec 5.3L	2005-2009	LH6	5.3 (325.2)	96.01 (3.780)	92.00 (3.622)	AL	AL	9.9:1	315	AFM (most versions)
	2007-2009	LY5	5.3 (325.2)	96.01 (3.780)	92.00 (3.622)	FE	AL	9.9:1	320	AFM
	2007-	LMG	5.3 (325.2)	96.01 (3.780)	92.00 (3.622)	FE	AL	9.9:1	326	AFM; Flex Fuel
	2007-	LC9	5.3 (325.2)	96.01 (3.780)	92.00 (3.622)	AL	AL	9.9:1	326	AFM; Flex Fuel; VVT (some versions)
	2008-2009	LH8	5.3 (325.2)	96.01 (3.780)	92.00 (3.622)	AL	AL	9.9:1	300	
	2008-	LMF	5.3 (325.2)	96.01 (3.780)	92.00 (3.622)	FE	AL	9.9:1	310	Flex Fuel; VVT (some versions)
	2010-	LH9	5.3 (325.2)	96.01 (3.780)	92.00 (3.622)	AL	AL	9.7:1	300	VVT
LS4	2005-2009	LS4	5.3 (325.2)	96.01 (3.780)	92.00 (3.622)	AL	AL	10.0:1	303	AFM; used in FWD cars
LS2	2005-2009	LS2	6.0 (364.1)	101.6 (4.000)	92.00 (3.622)	AL	AL	10.9:1	400	
Vortec 6.0L	2007-	LY6	6.0 (364.1)	101.6 (4.000)	92.00 (3.622)	FE	AL	9.6:1	360	VVT
	2007-2009	L76	6.0 (364.1)	101.6 (4.000)	92.00 (3.622)	AL	AL	9.6:1*	367	AFM; VVT (truck versions)
	2008-2009	LFA	6.0 (364.1)	101.6 (4.000)	92.00 (3.622)	AL	AL	10.8:1	332	"Hybrid, AFM,
VVT, LIVC										
	2010-	L96	6.0 (364.1)	101.6 (4.000)	92.00 (3.622)	FE	AL	9.6:1	360	Flex Fuel (some versions); VVT
	2010-	LZ1	6.0 (364.1)	101.6 (4.000)	92.00 (3.622)	AL	AL	10.8:1	332	Hybrid, AFM, VVT, LIVC
Vortec 6.2L	2007-2008	L92	6.2 (376.0)	103.25 (4.065)	92.00 (3.622)	AL	AL	10.5:1	403	VVT
	2009-	L9H	6.2 (376.0)	103.25 (4.065)	92.00 (3.622)	AL	AL	10.4:1	403	Flex Fuel; VVT
	2010-	L94	6.2 (376.0)	103.25 (4.065)	92.00 (3.622)	AL	AL	10.4:1	403	AFM; Flex Fuel; VVT
LS3	2008-	LS3	6.2 (376.0)	103.25 (4.065)	92.00 (3.622)	AL	AL	10.7:1	436	
LSA	2009-	LSA	6.2 (376.0)	103.25 (4.065)	92.00 (3.622)	AL	AL	9.0:1	556	Supercharged
LS9	2009-	LS9	6.2 (376.0)	103.25 (4.065)	92.00 (3.622)	AL	AL	9.1:1	638	Supercharged; dry sump oiling system
L99	2010-	L99	6.2 (376.0)	103.25 (4.065)	92.00 (3.622)	AL	AL	10.4:1	400	AFM; VVT
LS7	2006-	LS7	7.0 (427.6)	104.78 (4.125)	101.6 (4.000)	AL	AL	11.0:1	505	Dry sump oiling system

*10.4:1 on car versions.
"**Estimate; varies by vehicle and model year. Where appropriate, the highest horsepower rating the engine received as of MY 2010 is listed."

"Note: The above designations pertain to the United States market only, but the same or very similar engines are used in other North American markets as well as overseas."

The original Gen III, the LS1. All subsequent small-blocks draw their roots from this engine. Though features like a cam-in-block, pushrod architecture, and the 4.4-inch bore spacing of the original Gen I were retained, virtually nothing else was. This means that as good as the older small-blocks were, this new engine design offered reliability and power potential above and beyond those previously possible. (Illustration courtesy of General Motors)

effectiveness is definitely not limited to North America!

A detailed analysis of the ground-up design of the Gen III would fill volumes, and even a brief dabble in the literature on this engine inspires awe. We're not here to duplicate that information, but a complete gloss-over of the glory of Gen III engineering would be an injustice. That said, below are a few of the more major design advances GM incorporated into its new-generation small-block (we'll also go into more detail on many of them later in the book where they become relevant).

Aluminum engine block

The Gen III marked the first time an aluminum block was used by GM for a mass production pushrod V-8. Aluminum's main advantage is that it allows an engine block to weigh up to 50% less than a similar block cast from iron. But not all Gen III engine blocks were made of aluminum; while it's GM's material of choice for engines destined for passenger cars, trucks sometimes ended up with iron blocks (see our Gen III/IV Engine RPO Table for this information).

Cross-bolted main bearing caps

Most Gen I and II engines had only two bolts securing each main bearing cap to the block, while some higher-performance engines had four. By virtue of using a deep-skirt design for the Gen III engine block, GM was able to not only use 4 bolts holding each of the 5 caps to the block, but incorporate an additional 2 holding each from either side.

Revised firing order

The Gen III did away with the familiar 1-8-4-3-6-5-7-2 firing order of previous small-block engines. The new firing order (1-8-7-2-6-5-4-3) reduced crank arm stresses, quelled vibration, and improved main bearing performance.

Cathedral-port cylinder head

The radically different-looking cylinder head that debuted on the Gen III LS1 was designed in the interest of meeting many needs. The most striking feature is that its ports are tall and narrow. Shaped partially by the desire to target the fuel injectors at a specific location on the intake valve, this profile was also simply mandated by the narrow spacing afforded between the head bolts and pushrods. The ports are also replicated, so they're identical for each cylinder.

Revised valve angle

In the interest of engine durability, ease of manufacture, and other considerations, the Gen III retained the inline valve setup of previous small-blocks. However, its valve angle was changed from the traditional 23 degrees to 15 degrees from vertical. Among other things, this specification streamlines the transition from the exhaust port floor to the valve seat and creates a shallower combustion chamber. All of this translates to improved engine efficiency and output. This valve angle would be reduced even further on some Gen IV engines.

On a Gen III block, there are 6 bolts securing each main cap (including the 4 visible here and one on either side), for a total of 30 bolts holding the crankshaft in place. This stands in stark contrast to an old 2-bolt small-block's total of only 10!

The distinctive Gen III intake port opening. Later on, we'll explain how some Gen IV engines diverge from this port shape; but nonetheless, you should know that the engineering that went into Gen III cylinder heads yielded efficiency previously unheard of in a production small-block head.

Optimized valvetrain

Even though the Gen III retains a pushrod-actuated OHV design, all components were maximized for high stiffness and low moving mass. The Gen III also incorporates a so-called net build scenario for valve lash—this pretty much means that the valvetrain is nonadjustable. GM saw this approach as more robust, in part because with rocker assemblies rigidly bolted to the cylinder heads, valve lash would be less likely to change. Also, the new rocker arm design is significantly different from the traditional stud-mount units used in previous small-blocks. Their cast, rollerized design allows for better stiffness and less rotational inertia.

Composite intake manifold

Gen I small-blocks used various types of intake materials, and Gen II engines used aluminum exclusively. But for the Gen III project, GM invested in the development of thermoplastic intake manifolds. This was thought worthwhile thanks to lighter weight (the LS1 intake was less than half the weight of the LT1, a savings of over nine pounds), not to mention the new material's heat insulating characteristics helping to yield a cooler intake charge and reduced temperatures of fuel running through the fuel rails. In addition to the new material composition, the manifold no longer sealed the lifter area—a significant source of leaks over the years. Instead, a separate valley cover resulted in the manifold being totally isolated from the inside of the engine.

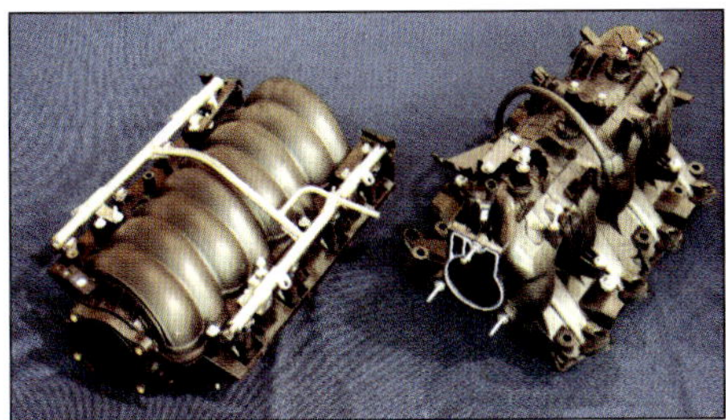

Thermoplastic was the material of choice for the LS1's intake manifold, and this was true for all other Gen III—and later, Gen IV—intake manifolds (except supercharged models). They would also come in many different shapes and sizes: this is a typical car intake (left) compared with a typical truck intake (right). As you can see, car intakes are lower in profile, which affords additional hood clearance.

Distributorless ignition

In order to deliver spark ignition, the Gen I used a rear-mounted distributor driven by a gear on the back of the cam, while the Gen II had the aforementioned front-mounted distributor, which was driven off a pin on the front of the cam. The Gen III has absolutely no distributor provisions, instead relying on eight individual computer-controlled coils to deliver spark. A crankshaft position sensor mounted at the rear of the block reads an encoded sensing ring (a.k.a. reluctor ring) located at the rear of the number eight counterweight—this is near the lowest deflection point in the crank. A separate camshaft position sensor serves to indicate which half of the firing sequence the engine is in. Simply put, this coil-near-plug system reduces losses inherent in a mechanical switching distributor and through long secondary leads. It's far more precise, and GM claims it resulted in a net ignition energy increase of 50 percent.

New sealing technology

Past small-blocks relied on a fairly wide variety of gasket material and RTV silicone to seal metal parts of the engine together, and they often had to adhere to curved surfaces and join parts that met at an angle. To help reduce the risk of leaks, the Gen III was designed with extensive use of single-plane sealing surfaces, and utilizes so-called controlled compression aluminum carrier gaskets that are a hybrid of silicone and aluminum.

Revised oil pump location

Gen I and II oil pumps hung on the rear main bearing cap and were driven off of the back of the camshaft. But for the Gen III, the oil pump is a gerotor design that sits on and drives off of the front of the crankshaft. This change alone reduced the block length by well over an inch compared to the previous-generation small-block, and it also allowed for the use of a shallower oil pan, resulting in more favorable engine packaging options.

With these and other improvements all working in conjunction, the Gen III took the automotive scene by storm. After winning accolade upon accolade from the press and industry alike, it quickly became the standard pushrod V-8 engine by which all others would be judged. The 5.7L LS1 spread to the Camaro and Firebird in 1998, and in 1999, GM began introducing truck variants in 4.8L, 5.3L, and 6.0L displacements. And as if the muscle provided by the LS1 weren't enough, GM was quick to up the ante with its 2001 model year release of the LS6, an engine whose output greatly eclipsed that of the LS1 thanks to improvements like a higher-lift cam and revised cylinder heads.

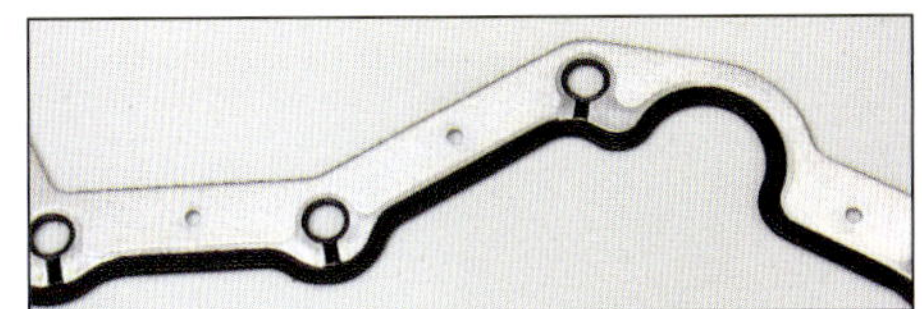

Silicone/aluminum carrier gaskets not only improve seal and compensate for fastener torque variation, they also allow the engine covers to become structural members. Even the valley cover beneath the intake manifold contributes to block stiffness! This type of gasket is also great news from a rebuild perspective because it virtually eliminates the use of goopy RTV silicone, making both disassembly and assembly far less messy.

Yes, the Gen III family was an official success—but never one to rest on its laurels, GM almost immediately began to finesse its already-stellar design, and soon enough the Gen III itself was to be superseded.

The Gen IV: Improvements Abound

Beginning in model year 2005, GM introduced the first of a slew of revised versions of its enormously successful new small-block, and these new additions to the product line carried the designation of Gen IV. Taken as a whole, the mechanical differences carried in this new generation of engines were minor—enough that parts compatibility between Gen III and Gen IV engines is more than substantial. At the same time, though, the Gen IV began incorporating many impressive new technologies rarely or

You can find engines based on the Gen III/IV small-block architecture under the hoods of many GM vehicles, including sports cars, muscle cars, vans, trucks, and SUVs—proving not only that this family of engines offers impressive horsepower, but that it's capable of meeting the extreme durability requirements of heavy towing and hauling.

The LS1 was not the lone Gen III in the stable for long. Beginning in 1999, truck variants began appearing, like this 6.0L RPO LQ4. Though they share the same basic engine architecture as car versions, Gen III engines designed for light truck duty are easily distinguishable by their taller intake manifolds. (Photo courtesy of General Motors)

never before seen on cam-in-block engines.

One of the Gen IV engine program's main revisions included moving the camshaft position sensor from the upper rear of the block (where it read off the back of the camshaft) to the front cover. Additionally, the location of the knock sensors was changed from the lifter valley area to the exterior lower sides of the block. The primary reason for the relocation of these items was to accommodate GM's exciting new Displacement-On-Demand (DOD) technology, also known as Active Fuel Management (AFM). By deactivating half of the engine's cylinders under certain light load conditions, AFM provides significant fuel savings. This is all accomplished via special switching valve lifters and a so-called Lifter Oil Manifold Assembly (LOMA) located in the lifter valley. But, for reasons mainly involving the particular vehicle application of each engine, not all Gen IV V-8s had the AFM system. You can see in the accompanying Gen III/IV Engine RPO Table a rundown of which engines featured AFM, and surely more such variants are being released as you read this. Also debuting on some members of the Gen IV line

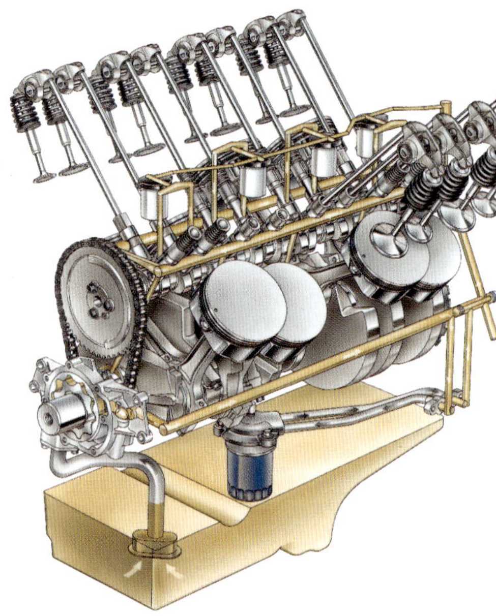

First introduced on some Gen IV engine variants in the 2005 model year, GM's Displacement-On-Demand (a.k.a. Active Fuel Management) system does more than simply cut fuel to half of the cylinders. This diagram shows the oil flow routes that, on command, trigger special lifters to temporarily stop transmission of the cam lobe profile to their pushrods. With the valves closed on deactivated cylinders, pumping losses are virtually eliminated. Claimed fuel efficiency increases of 8% were enabled with the introduction of AFM. By the way, the use of electronically controlled throttle bodies was now across the board for the Gen IV, and part of the reason is because they're needed for seamless, imperceptible switching between 4- and 8-cylinder mode on AFM-equipped engines. (Illustration courtesy of General Motors)

One of the first Gen IVs, the 5.3L LH6. Though some noteworthy changes were made to the design, the Gen IV is essentially the same engine as the Gen III, and it was probably more in the minds of GM's business brass than in those of the engineering folks that the changes merited a new generation name. All things considered, the difference between the two is far milder than even the Gen I to Gen II distinction. (Illustration courtesy of General Motors)

were features like E85 Flex-Fuel capability, variable valve timing (VVT), late intake valve closure (LIVC), and even hybrid gasoline/electric drive systems. Though these and other impressive technological features were unique to the Gen IV, some Gen IV engines failed to incorporate a single one of these advances—so judging the break between the Gen III/IV is not as simple as it may seem.

Other Gen IV features were simply carryovers of late-Gen III improvements or extensions of them over a broader range of engine offerings. These included items like an improved timing chain, now incorporating a vibration dampener (a feature originally developed for the LS1 but that hadn't made the production cut) or even a tensioner on some applications. Also taking a cue from some later-production and high-performance Gen III variants, all engines now featured floating piston pins and coated piston skirts for reduced noise and increased durability. Other mild upgrades like revised coil packs separate a Gen IV from its immediate predecessor, but just as one example of the two's similarity, Gen III LS6 and Gen IV LS2 engines utilized the exact same cylinder head casting!

In sum, these differences taken as a whole are enough to distinguish Gen III and Gen IV block castings and make them technically incompatible with one another, even though the majority of engine parts are interchangeable. (We should also note that, with some help from the automotive aftermarket, it's usually a straightforward task to convert from one style of engine to the other—see our "Block-Swapping Points of Interest" Workbench Tip in Chapter 4.) Again, we'll cover the Gen III/IV distinctions in detail where they become important in later chapters—but as far as we're concerned, they are two similar versions of the same spectacular LS engine!

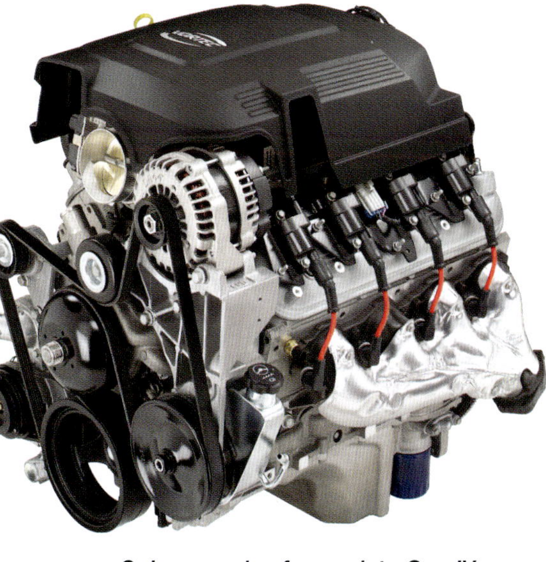

Only a couple of years into Gen IV production, some very serious high-output truck engines began to join the lineup, a perfect example being the 6.2L RPO L92. Though this particular engine did not feature AFM or Flex-Fuel capability, it did incorporate VVT. It was also one of the first truck engines to include a revised cylinder head design with greatly improved flow capability (Also see "LS Cylinder Head Evolution," on page 57). (Photo courtesy of General Motors)

One of the most impressive GM engines ever released, the Gen IV 7.0L LS7 that debuted in the 2006 Corvette Z06 featured 505 hp along with titanium connecting rods, CNC-machined cylinder heads, and a race-ready dry-sump oiling system. Hand built by a highly trained team at GM's Performance Build Center, the LS7 was a work of mechanical majesty, and though arguably more impressive LS engines were still to come, it represented the epitome of overhead-valve high-performance engine technology the moment it was introduced. (Photo courtesy of General Motors)

This cutaway view of a GM display engine illustrates another new technology introduced on some Gen IV engines: Variable Valve Timing (VVT), also known as variable cam phasing. By use of a vane-type phaser mounted to the camshaft sprocket, the cam can be turned relative to the crankshaft, delivering optimally timed valve events at any engine speed. Though the system cannot independently adjust intake and exhaust timing (unlike similar systems on DOHC engines), it still allows for increased performance and efficiency.

Determining Whether Your Engine is a Gen III/IV

We've created a couple of helpful tables in this chapter to help you figure out exactly which LS engine you've got in your hands. One lists all variants of the Gen III/IV, along with their RPO designations and some important specifications. The other lists all GM passenger cars equipped with Gen III or IV engines. (Because there are so many more versions of trucks and truck engines offered by GM, and because of the aforementioned Gen I and III overlap, it's best to simply refer to the RPO table for exactly what engine your truck has.) Combine these tables with the other information in this chapter, and you shouldn't have a problem determining whether your engine is an LS.

The Small-Block Lives!

We hope that even seasoned LS fans have learned something from this chapter about the Gen III/IV small-block and its roots. But before ever opening this book, any reader probably knew that the GM small-block has always been about producing gobs of power and torque out of a sensible package: whether it be for speed, towing capacity, or just that extra oomph to make driving more pleasurable. Perhaps you'd like to rebuild your old Gen III or IV so that it will perform like new, or maybe you're here to unlock additional power out of your LS (we'll begin our next chapter with a discussion of these goals). Whether you're here for one reason or the other—or are even building an LS from scratch—we've got you covered. Welcome!

Gen III/IV-Equipped U.S. Passenger Cars

Make	Model	Submodel	Years	Engine	Generation
Buick	LaCrosse	Super	2008–2009	LS4	Gen IV
Cadillac	CTS	CTS-V	2004–2005	LS6	Gen III
			2006–2007	LS2	Gen IV
			2009–	LSA	Gen IV
Chevrolet	Camaro	Z28, SS	1998–2002	LS1	Gen III
		SS	2010–	LS3	Gen IV
			2010–	L99	Gen IV
Chevrolet	Corvette		-1997–2004	LS1	Gen III
		Z06	2001–2004	LS6	Gen III
			-2005–2007	LS2	Gen IV
		Z06	2006–	LS7	Gen IV
			-2008–	LS3	Gen IV
		ZR1	2009–	LS9	Gen IV
Chevrolet	Impala	SS	2006–2009	LS4	Gen IV
Chevrolet	Monte Carlo	SS	2006–2007	LS4	Gen IV
Pontiac	Firebird	Formula, T/A	1998–2002	LS1	Gen III
Pontiac	G8	GT	2008–2009	L76	Gen IV
		GXP	2009	LS3	Gen IV
Pontiac		GTO	-2004	LS1	Gen III
			2005–2006	LS2	Gen IV
Pontiac	Grand Prix	GTP	2005–2008	LS4	Gen IV

Note: GM is constantly expanding not only its line of small-block V-8s, but the passenger cars they are installed in—so there's little doubt that more are being introduced as you read this!

Still Confused?

Maybe you have an engine sitting in your garage that you don't know the exact vehicle it's from, or even the model year it was produced. In that case, information on vehicle applications and RPO designations is probably not very helpful. Therefore, let's reiterate some easy, hard-and-fast rules to help rule out other engines:

No overhead cams

If you pop a valve cover on your V-8 and you've got a cam or two sitting under there, it's not a Gen III/IV.

No distributors

Look at the ignition system. If you see a distributor–either in the back or in the front–it's not a Gen III/IV.

No metal intakes

Aluminum or other metal intake manifolds were never installed on any Gen III or IV engines (except supercharged versions). Unless the engine in question has an aftermarket intake installed, it's not a Gen III/IV.

No iron heads

Look closely at, or better yet, take a magnet to your engine's cylinder heads. If they're magnetic, they're iron, and therefore it's not a Gen III/IV. The sole exception to this is the early 6.0L RPO LQ4 truck engine, which had iron heads for its first couple of years of production.

These final notes should help eliminate any residual confusion you may harbor about Gen III/IV engine identification.

Engine Rebuilding: Skills, Techniques and Tools

Now that you have a better idea of what Gen III and IV LS small-blocks are and whether the engine you're looking to rebuild falls within the family, it's time to decide whether an engine rebuild is really something that you want, or need, to do. To determine this, you first need to know what, exactly, we mean when we use the term "rebuild." After this, there are some important questions you'll want to ask, like: "Is this something I am capable of?" "What tools are needed?" And, "Where can a rebuild be performed?" We'll try to answer all of these and more in this chapter.

What is a "Rebuild?"

The internal combustion engine in the form we know best today began widespread use in automobiles and other consumer products around the turn of the 20th century. The four-stroke Otto-cycle internal combustion engine variation (as

Rebuilding for High-Performance?

The first part of this chapter is a general overview of what an engine rebuild is, the very basics of what it entails, and the inevitability of such an overhaul eventually being necessary on every engine ever manufactured. None of these explanations are needed if you've decided to rebuild your engine for high-performance.

While most people think of hopping up an engine as modifying it for "performance," this term can be confusing as it technically refers to the myriad general concerns that apply to any engine. In a nutshell, an engine performs well if it has the ability to efficiently and reliably turn the chemical energy stored in fuel into motion down the road.

On the other hand, high-performance entails the likes of tire-smoking horsepower for street and racecars, or serious towing and hauling torque for trucks, vans, and SUVs. Anyone in pursuit of these goals probably already knows that a full engine overhaul is the best possible way to unlock additional power (along with the reliability needed to sustain it). The great news is that it's possible to add substantial horsepower and torque to an LS on just about any budget: thanks to the advanced engineering incorporated into the Gen III and IV, these engines respond well even to mild internal modifications. And for the less cost-conscious rebuilder, the aftermarket is awash in quality components needed to sustain even higher power levels! We'll go over how to accommodate tried-and-true techniques like increased cubic inches, ported cylinder heads, and high-lift camshafts into your rebuild as we go along... for now, just know that if you're rebuilding for high-performance, any concerns over whether your engine is really worn-out enough to merit a rebuild don't apply to you–and neither does the first part of this chapter!

Socket wrenches: You will need a variety of regular and deep socket wrenches, both 12-point and hex, and they'll need to be metric because that's what everything on an LS is! You should have standard sockets as well, though, particularly if you'll be installing any aftermarket parts (which many times use standard bolts and nuts).

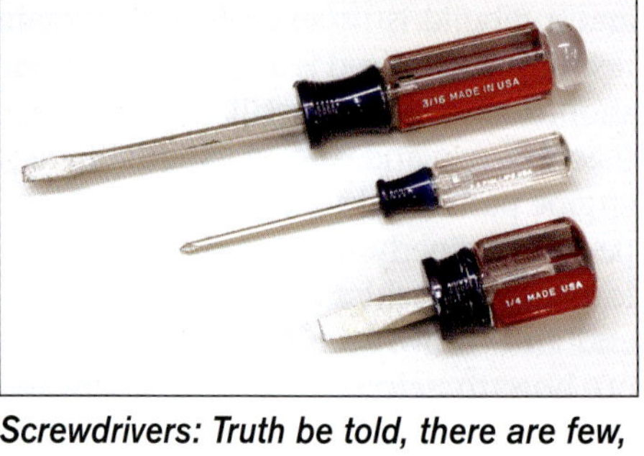

Screwdrivers: Truth be told, there are few, if any, screws on an LS engine; but you'll still need screwdrivers and will be surprised how often they come in handy, especially flatheads.

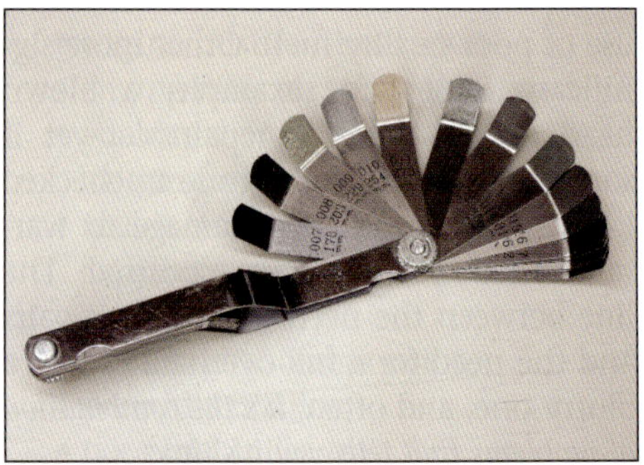

Feeler gauge: Used primarily during ring fitting, a feeler gauge will also be useful for other measurements, like checking engine cover alignment.

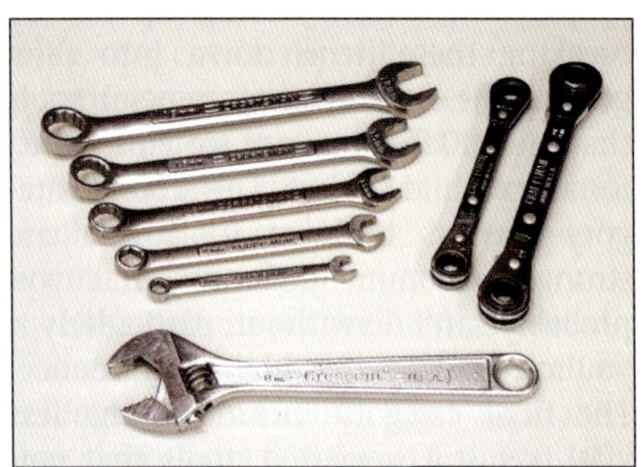

Open- and box-end wrenches / pliers: Open-end and box-end wrenches should also be metric, though standard sizes can be useful too. Have a few adjustable wrenches ready as well. Ratcheting box-end wrenches can be a time saver. Not shown but helpful are pliers, both regular and needle-nose.

Magnets: During disassembly, magnets will be a big timesaver and are particularly helpful at removing slippery, oily items like lifters. You also will need one when you manage to drop a bolt into a hard-to-reach place (notice that is a "when," not an "if!").

Piston ring compressor: Don't just buy any old ring compressor that looks like it should be an oil filter wrench. Tapered, sleeve-style ring compressors are the preferred type as they allow the best chance of your piston rings safely entering the bore without breakage. Some are fixed for a very narrow range of bore sizes (right), while others are adjustable over a small range in bore size (left).

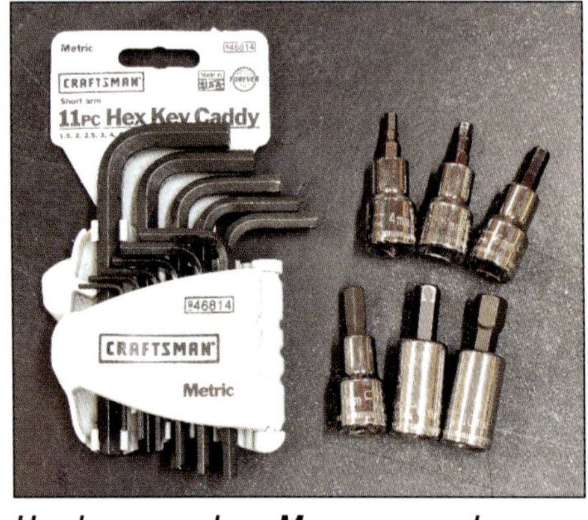

Hex key wrenches: More commonly known as "Allen" wrenches, you will need metric versions of these in certain places on an LS engine, one example being the oil gallery plugs. Some engines have star head (TORX) bolts as well, so have a set of those around too.

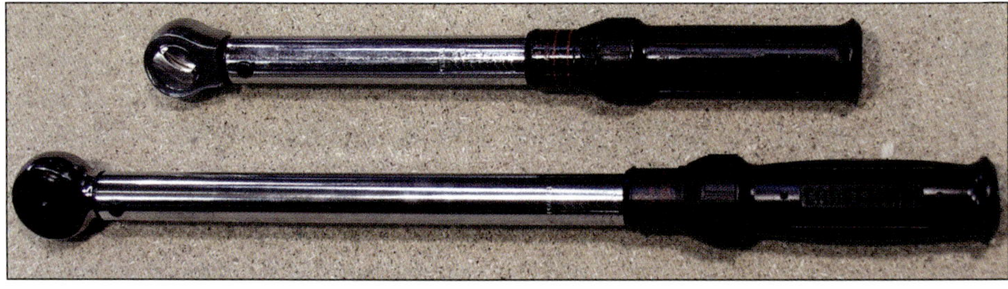

Torque wrenches: You will need at least two of these: one that works in inch-pound increments, and another for low-to-medium foot-pound readings. You may need a third wrench that is good to high foot-pound readings (75 ft-lbs or more), especially for a high-performance rebuild using aftermarket fasteners. Make sure your torque wrenches are fairly new—these things lose their accuracy over time! If there is any doubt, you may want to take them back for calibration (many tool outlets do this).

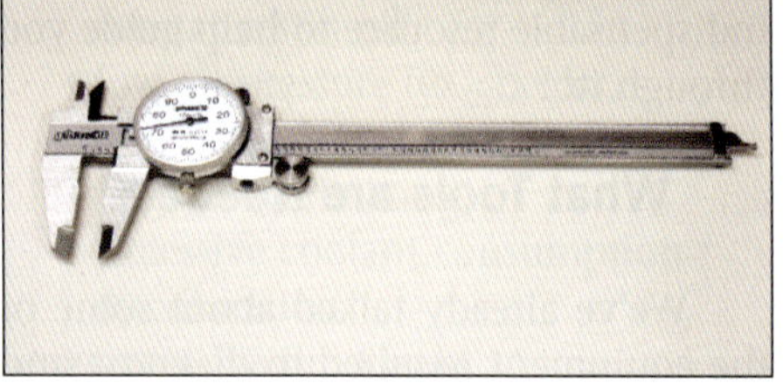

Dial caliper: Calipers are convenient for taking measurements of a variety of parts, whether you need to know an item's diameter, length, or depth. Most calipers have a dial readout, but they are also available with electronic readouts for even better accuracy.

Goodson Tools and Supplies for Engine Builders

While local hardware stores can be decent sources for some of the more basic tools you'll need for an engine rebuild, attempts at securing more specialized items like piston ring filers and precision measuring tools from them would be largely unsuccessful. You could drive around for days searching every possible tool source in your area, and still not come up with half of the items you require. Fortunately, there are dedicated engine tool and supply outfits out there that stock just about everything you will need for a rebuild, and one of the best is Goodson.

This virtual one-stop source sells thousands of products aimed at both professional engine machinists and do-it-yourself engine assemblers alike. Even a cursory glance through the latest Goodson catalog will make you wonder if there's anything the company doesn't stock. From piston ring compressors to valve spring height gauges, if Goodson doesn't have it, it probably doesn't exist! Check out the company website at www.goodson.com, or call 800-533-8010 to get a copy of the latest catalog.

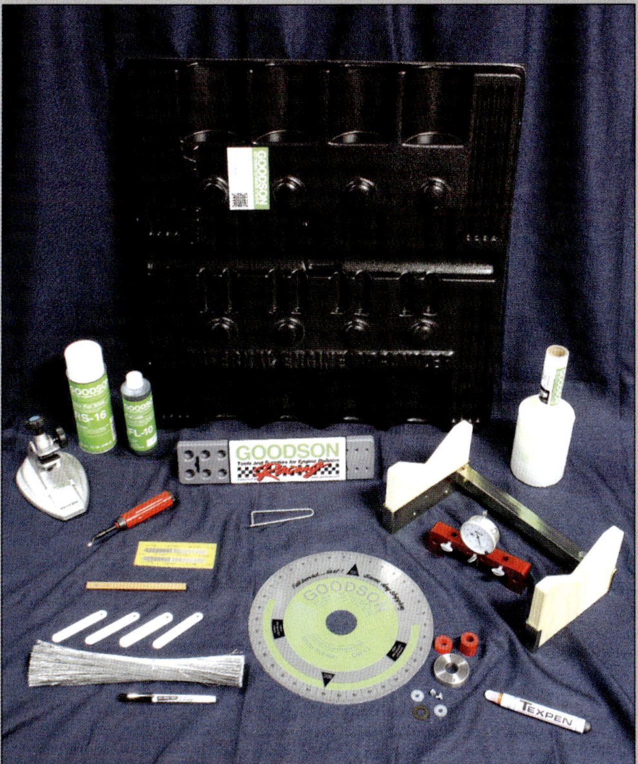

Goodson markets a wide variety of tools and supplies, and what's seen here is just the tiniest sampling of it. Whether you're looking for organization items like parts ID tags, high-performance tools like degree wheel kits, or anything in between, Goodson has it, and is continually expanding its product line.

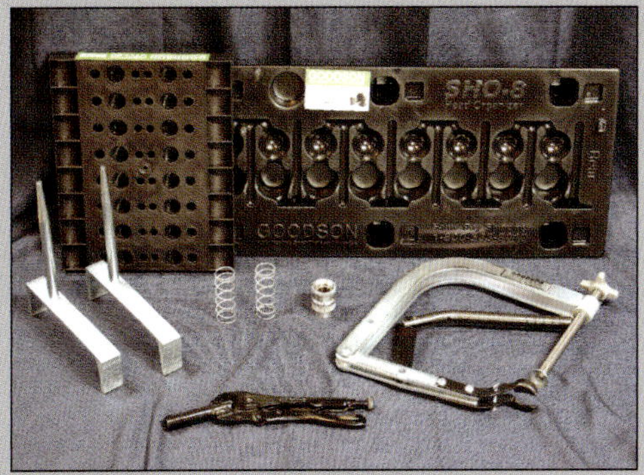

Here's just a smattering of the tools Goodson sells that will be helpful for working on the cylinder heads and valvetrain of your LS. The black items in the back are organizer trays—why not keep your pushrods and valves in a secure place, easy at hand for when the time comes for assembly (not to mention serving as helpful storage during disassembly)? Organization is of utmost importance in a rebuild!

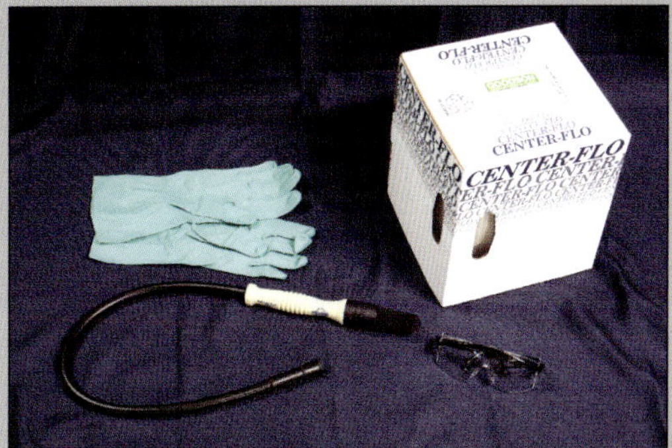

These are just a few of the cleaning supplies you can order from Goodson, including protective gloves, assembly wipes, goggles, and a parts washing brush. We'll talk more about component cleaning and the supplies needed for it in Chapter 6.

We explain in the main text that a sleeve-style ring compressor works best, and Goodson sells them. Still, this adjustable unit from Goodson has the advantages of being less expensive and having the ability to be reused during other rebuilds (on engines of widely different bore sizes). This unit also includes a squaring tool and is just an example of the variety of tools Goodson has to offer, no matter what your needs or budget.

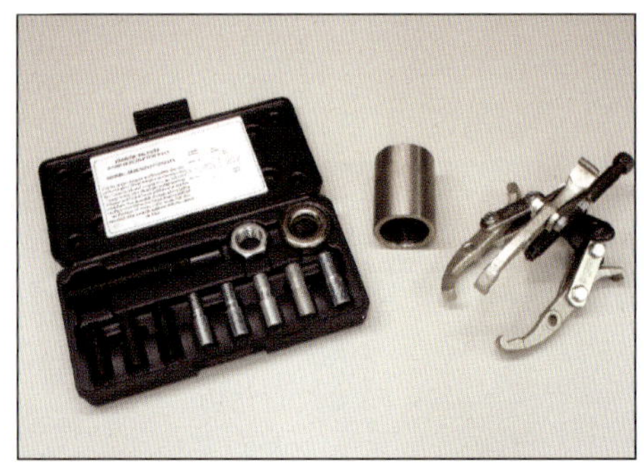

Harmonic damper and gear pullers/installers: As with past small-blocks, the LS's harmonic damper (or "balancer") requires a harmonic balancer installation tool (left) to press it onto the snout of the crank. You must ensure, however, that the tool you pick up is compatible with M16 x 2.0 threads and that it has sufficient reach to get at the deep-set threads in an LS's crank snout. If not, a backup plan is available that will simply involve buying a longer bolt (see Chapter 8 for details). Along similar lines, although the gears of some timing sets slide easily onto the crank snout chances are you will need a crank gear installing sleeve (center) to help the job—though there is a backup plan here, too, that involves using your old crank sprocket to "double up" and press on the new one (see Chapter 8 as well). A 3-jaw puller (right) will assist in removing both of these items during disassembly.

Engine stand and hoist: If you've been working on cars and engines for any amount of time, you already have these. You'll be doing nearly every step of the rebuild on an engine stand, while a hoist will be needed for getting the engine in and out of the vehicle and/or its removable subframe. They don't have to be shiny and new, so long as they are in good condition and fully functional!

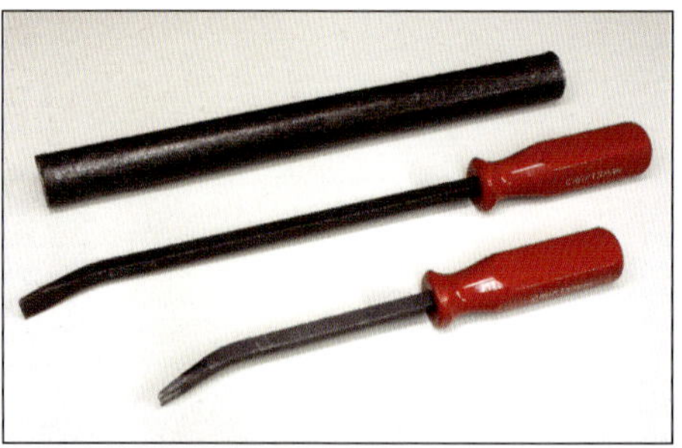

"Cheater" bars/pry bars: These will aid primarily in engine disassembly, but may be helpful during some assembly steps as well. Not pictured but very useful: a 1/2-inch drive breaker bar.

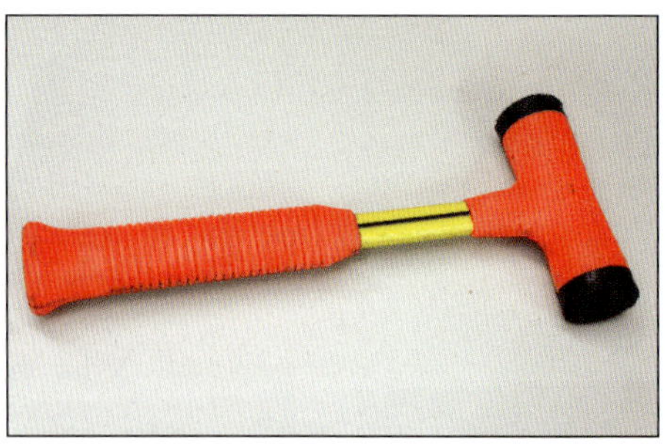

Rubber mallet: Despite what you're thinking, this is not just for disassembly! A rubber-faced mallet will be used during assembly as well. A standard metal one will be helpful too, though for more limited purposes.

Recommended Tools

These tools are very strongly suggested, and though you theoretically could put an engine together without them, it would be a very bad idea! Some of them actually *are* mandatory, but have been placed in this category simply because an alternative tool exists that you may substitute if desired. A select few will likely only be needed in high-performance rebuilds, and these will be noted accordingly. A choice not to invest in nearly all of these would be a grave mistake!

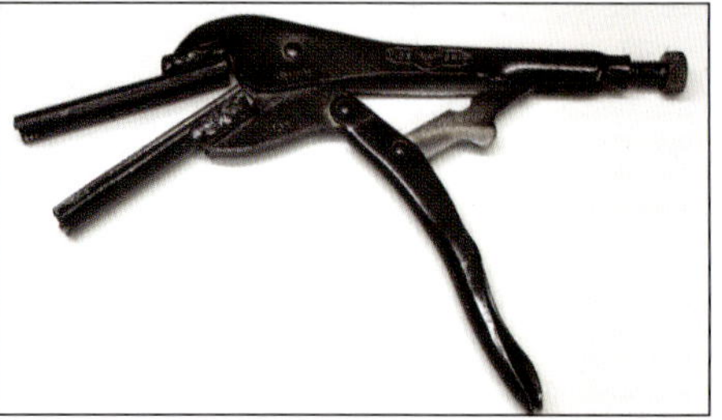

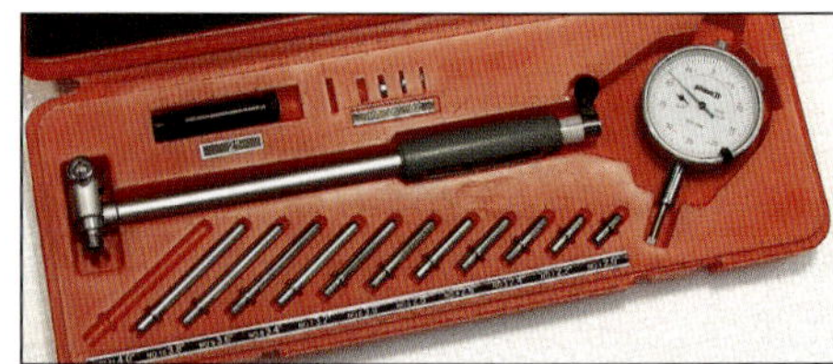

Dial bore gauge: This type of gauge is useful for accurately measuring diameters of cylinders and bearing bores. It primarily will help you in verifying the quality of machine work after it's been performed.

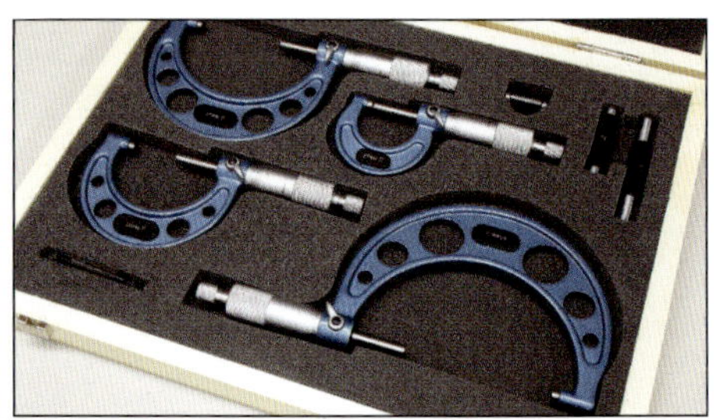

Micrometers: A set of micrometers is relatively inexpensive, though price rises with dimensional accuracy. Since you really only need a set good enough for double-checking diameters of machined items like journals and piston pins, you should grab a set accurate to a half thousandth of an inch (0.0005).

Degree wheel kit: Normally needed on high-performance applications only, a degree wheel kit allows precise verification and/or adjustment of camshaft timing during pre-assembly. A system for mounting a dial indicator and measuring piston and lifter movement is a required complement to a degree wheel kit, and some kits include it.

Valve seal puller: To ease the task of removing your valve seals from your cylinder heads (if you'll be disassembling yourself), a valve seal puller tool is a nice item to have. Its geometry makes providing the correct crush pressure to dislodge the seal from the top of the guide easy.

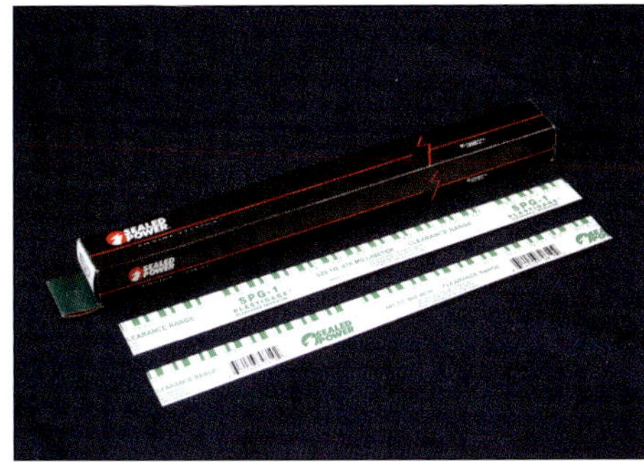

Gauging plastic: Whether you want to call it a supply or a disposable tool, this stuff is the easiest and most foolproof way to measure bearing clearances. It's most commonly known as Plastigage, a trade name used by Sealed Power. The only reason we're not calling it mandatory is that you theoretically could use a micrometer, a caliper, and a dial bore tool for the same job—but you'd need very accurate versions of these tools to do it properly.

Piston ring filer: This is for use when fitting piston rings to the bores of your Gen III/IV block. A hand-crank version works just fine at setting the end gaps, though electric versions are available that are more accurate and will save some time. The one and only time you will not need one of these is when using a "drop-in" ring set (even so-called gapless rings can require file fitting!).

Connecting rod vise: A standard vise is a bad way to hold rods secure while their bolts are being loosened. Unless you can find some specialized vice inserts that won't nick your rods, an aluminum rod vise is the best way to go, as they are relatively inexpensive. Steel rod vises with super-smooth clamping surfaces are also available, but can be very pricey.

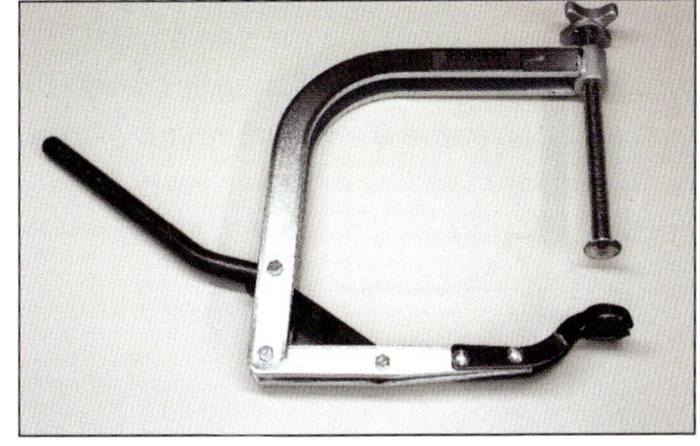

Valvespring compressor: This tool is used during both disassembly and assembly of cylinder heads, and you'll need one heavy-duty enough for automotive valvesprings. The only time you won't need a valvespring compressor tool is if you're having your machine shop disassemble and assemble your heads for you, or if you're swapping to a pre-assembled set of high-performance cylinder heads. The best type of tool locks in place when the spring is compressed, as this one from Goodson does.

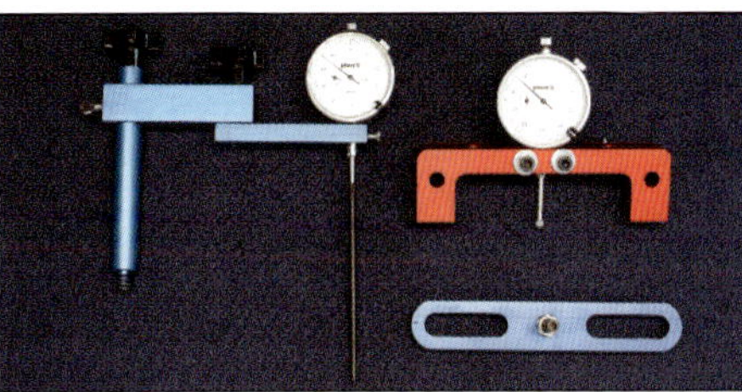

Dial indicator(s) and fixture(s): A dial indicator is an indispensable tool for measuring precise movement of many engine parts during pre-assembly. The fixture on the left is an adjustable one that will also prove useful for taking measurements like crankshaft endplay and runout, processes that are important on all engines. Try to get one designed for metric thread bolt holes if at all possible. The deck bridge-style fixture on the right is also helpful, but primarily for measuring piston movement and deck clearance. Shown in the bottom right is a TDC stop, helpful during cam degreeing (you'll need one capable of spanning the wide head bolt spacing of the LS).

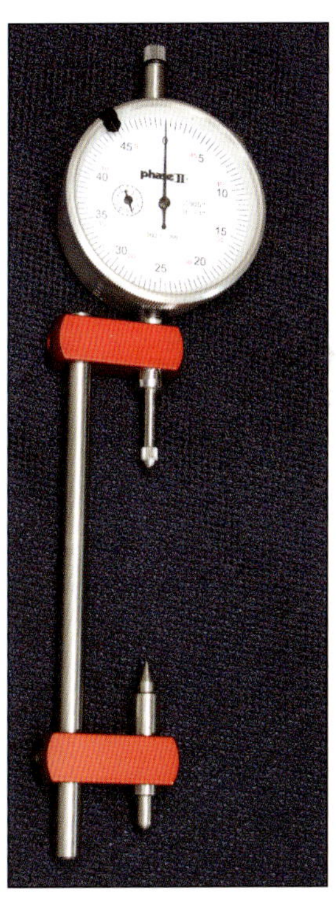

Rod bolt stretch gauge: Some high-strength aftermarket connecting rods use a bolt stretch specification for fastener tightening, and a gauge like this will be needed to measure it. If your connecting rod manufacturer doesn't specify this method, you won't need one of these.

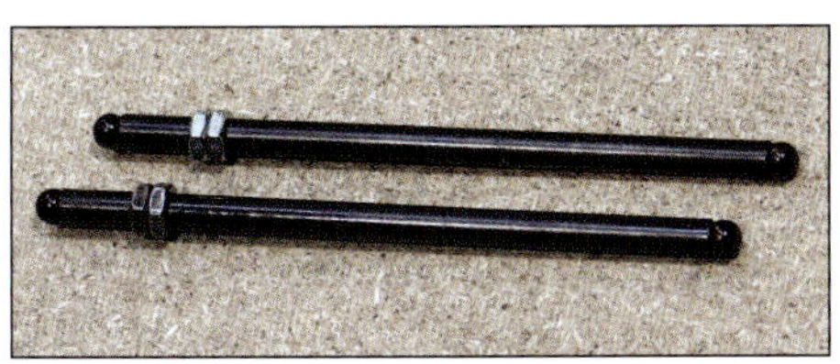

Pushrod length checker: Mandatory for most high-performance applications but also helpful for many stock rebuilds. It's best to have two checkers (one for each valve of a cylinder), and they'll need to work within the range of 7.400 inches (all LS engines have a stock pushrod length close to this). The best checkers are those strong enough to withstand the full load of a valve spring, so that you can use them during pre-assembly while turning the engine over and verifying things like rocker geometry.

Valvetrain checker springs: These items are needed for high-performance rebuilds only, and will be used during pre-assembly for proper verification of piston-to-valve clearance. You may want to wait until you have your valvespring retainers in hand to buy these–you'll need to match the checker springs to their O.D.

Valvespring height gauge: In order to ensure your valvesprings will be at the correct installed height (and thereby provide the right amount of seat pressure), you'll need a valvespring height gauge. While primarily used during high-performance rebuilds, this item can be useful in stock rebuilds, too. You will need one that will work in the range of your valvespring's required installed height and will fit the diameter of your valvespring retainers, so you should probably hold off on buying this gauge until you have those items.

Torque angle gauge: Some of GM's fastener tightening specifications used on the LS involve the torque-plus-angle method (for example, 15 ft-lbs plus 51 degrees–that's not something you want to "eye up"!). Unless you're using aftermarket fasteners exclusively, you will need a torque angle gauge.

Suggested Tools

Here are some other tools you may wish to get a hold of, depending on your application. Many of them are designed to help make your life easier by saving you some time and effort, reducing the chance for assembly (or disassembly) error.

Crankshaft rear oil seal installation tool: It is possible to gently tap the rear crank seal into the rear cover, but you can guarantee precise installation and a leak-free seal if you have the right tool. This one is made by famed LS engine specialists Wheel to Wheel Powertrain. While the two black pieces are what actually squeeze the rear seal into place, the aluminum donut also helps prevent the lips of some seals from being damaged as the cover is slid onto the back of the crank.

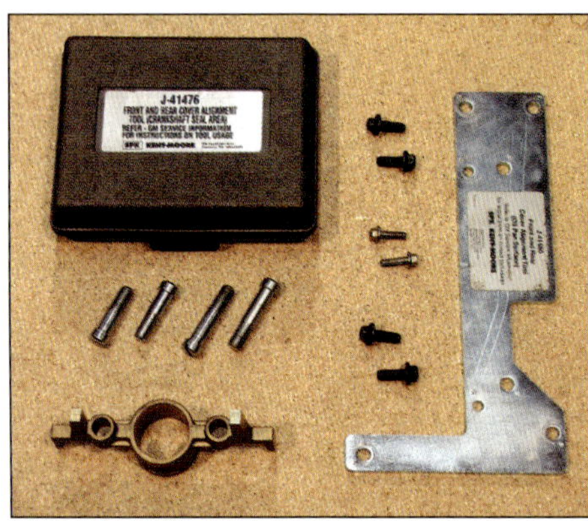

Engine cover alignment tools: The LS's front and rear engine covers require precise alignment with the block during installation. You can do this without special tools if you are very careful; however, it is more foolproof if you have the right equipment for the job. Made by SPX Kent-Moore, these are the official GM "J" tools, and they are very expensive, so try to find a tool outlet that will rent them to you!

Flywheel holding tool: There are a few times where you'll need to hold the crank steady in one position, and the best way to do this is to install your flywheel/flexplate on the back and use a flywheel holding tool (the one pictured is the official GM "J" tool). This item will be particularly useful during checking of rod bearing clearances, where any movement of the crank can destroy a measurement. For other steps, you can just use a suitable pry bar inserted into a flywheel tooth, but this is a little trickier.

Inspection stands: Having your heads lying on the surface of a table makes the processes of disassembly, inspection, and assembly somewhat more time-consuming. Similarly, your cam and crankshaft are bulky items that can be difficult to inspect and clean safely. Save yourself some time and frustration by picking up a set of head stands and a crank/cam inspection stand. Both of these are sold by Goodson and are good buys.

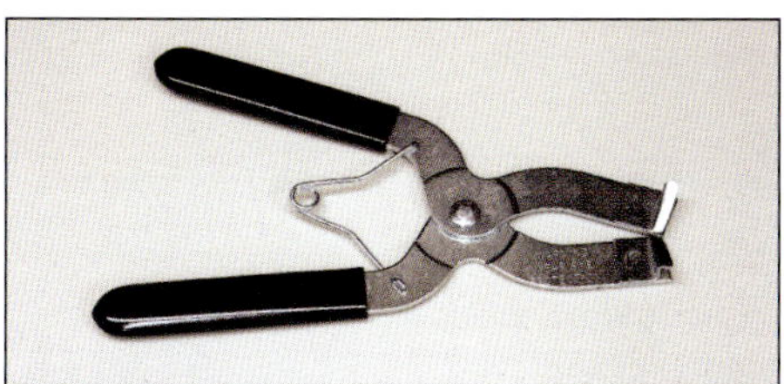

Piston ring expander pliers: These handy items will help you install the compression rings onto your pistons with reduced risk of ring breakage or piston scratching.

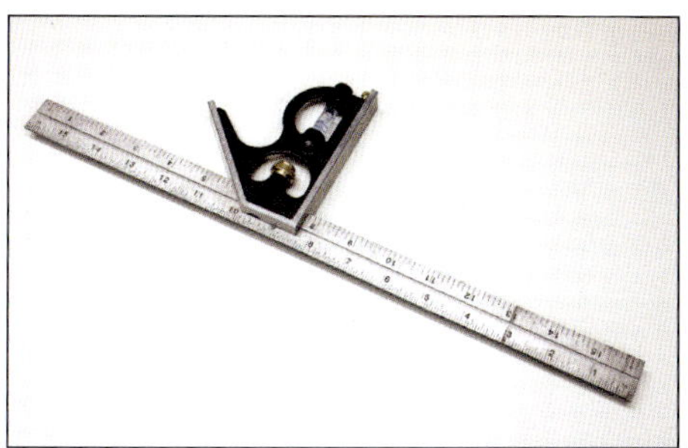

Straight edge: A super-flat machinist's straight edge isn't necessary because you'll be having your machine shop check deck flatness for you. However, a run-of-the-mill woodworking or similar straight edge can be helpful to check engine cover alignment.

Crankshaft turning socket: An engine build involves a substantial amount of spinning the crank, whether during disassembly, pre-assembly, or final assembly. For this reason, it makes things easier if you have a dedicated crank socket. LS-specific models are designed to fit onto the oil pump drive gear on the crank snout. Some, like this one made by COMP, even have a provision for installing a degree wheel onto the front. If you don't feel like spending the money on one of these, it is possible to do the job using an old crank bolt (noting that this will make it extremely difficult to turn the crank counterclockwise).

Where Can a Rebuild be Performed?

You've decided you're going to go ahead with rebuilding your LS engine and are confident that it's something you have the time and patience to undertake. You've also picked out the tools you are going to need. The pieces of the puzzle are all falling together... now, all you need is a place to do it.

Most of us don't have a dedicated, super-clean engine build room complete with climate control, large workbenches, and excellent lighting. This would certainly be nice to have, but fortunately you can get by with much less. For most do-it-yourselfers, this means a sufficiently large garage that's relatively clean and is isolated from intruders who might interfere with your work while you're away (or overly distract you while you're in the middle of it). Nothing is worse than stopping at a critical point in an engine assembly, only to come back a few hours later and find that someone moved the piston rings you've just uniquely file-fit to each cylinder, mixing them up in the process.

You will similarly want to avoid having anyone, for example, perform woodworking or other contaminate-creating tasks in the area while your project is underway (again, we'll talk more on the importance of cleanliness in Chapter 6). And let's state the obvious—alcohol and engine building do not mix, nor do distractions like watching a hockey game. An engine rebuild can be a fun project, but it's also one that requires a lot of care and attention to detail!

Conclusion

Now that you've got a better understanding of just what a rebuild is, and are better aware of the resources and tools you need to successfully complete it, you can move on to disassembling your LS engine. We'll go through this step-by-step in Chapter 3.

Fasteners and Threads: Problems and Solutions

While you may be able to perform an entire rebuild without running into a single fastener-related issue, there is always the chance you'll encounter a couple of stripped threads or stuck bolts. That's why we wanted to draw special attention to the issue of problematic fasteners, thread repair, and the tools and supplies you will need to deal with these problems.

The very best thing to do, of course, is to prevent thread damage from occurring in the first place. Often, damage occurs while attempting to remove a stuck or stubborn bolt during engine disassembly. Simply applying brute force is probably the worst thing you can do, as while the bolt may come out this way, it will often bring the hole's threads with it. To help prevent this from happening, be sure to spray ample liquid penetrant onto the threads of the bolt as soon as you begin to feel any unexpected resistance to movement. Allow it time to find its way down to the problem area in the hole and begin to break down any corrosion or hardened sealing compounds that are creating the problem. Even if metal-to-metal mechanical thread interference is to blame (for example, the bolt was overtightened when first installed), the penetrant will act to lubricate and minimize thread harm as the bolt is removed. In some cases, you may find that applying heat to the affected area can help loosen a stuck fastener; however, this should be done with caution, not only because of personal safety concerns, but since too much heat can damage engine components.

Despite your best efforts at a clean removal, thread damage can still occur as a bolt is being removed. In this scenario, you should have a tap set handy to attempt to get the

Fasteners and Threads: Problems and Solutions

CONTINUED

threads back into a usable form. Worse yet, a fastener may break and leave little or no surface to grab on to for any further removal attempts. Fortunately, bolt extractor kits are widely available to aid you in removing the remnants of the fastener. These kits often will do the job, but sometimes the extractor tools themselves will snap if a fastener is particularly jammed. In this case, your only choice is to drill the hole out completely and replace the threads.

Thread repair kits (often simply known by the brand name Heli-Coil) are great because when used correctly, the new threads are often stronger than the old threads ever were! (As proof of this, you may notice "helicoiled" threads on some aftermarket parts–for example, the rocker arm bolt holes of some aftermarket aluminum cylinder heads). Aside from replacing the threads of holes that fasteners have broken off in, these kits are also useful when threads are simply too damaged to be cleaned up with a tap. However, some patience and skill is required to use them, and you might consider having your machine shop perform these repairs for you. For example, a drill press or other specialized equipment may be required to make a hole of the proper straightness.

Sometimes, a thread problem does not arise until later on, during engine assembly. To prevent this, you should meticulously inspect all threaded holes in the block and other components before beginning assembly (as shown in Chapter 6, where we'll also demonstrate the usefulness of specialized clean-out taps). Never attempt to install a fastener into a hole whose threads look damaged or that have gunk buildup, and the same goes for the fastener!

This Workbench Tip is designed to help you not only deal with problems like this, but to prevent them from happening in the first place!

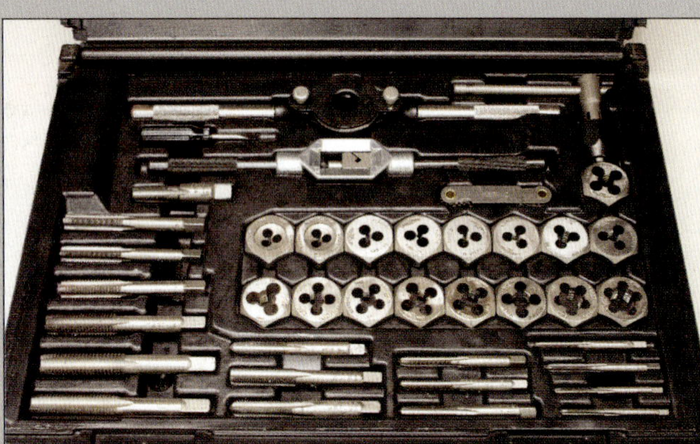

Minor thread damage can often be repaired via use of a tap set. Similarly, dies can be used on the threads of bolts that not only have suffered thread damage, but that may have leftover factory sealant that will prevent proper reinstallation. The best deal is to buy a full tap and die set–make sure it's a metric one! Here's an LS exhaust manifold bolt, and as you can see, buildup on them is notoriously bad.

Both Liquid Wrench and WD-40 are excellent penetrants that you should have at hand, ready to work their magic on any stubborn fastener.

Bolt extractor bits (left) are used for inserting into a broken bolt that has little or no area to grip onto. Kits are also available for bolts that have not actually broken, but have rounded-off heads that can't be easily gripped by normal tools (right).

Thread repair kits are widely available. Though they're designed with the do-it-yourselfer in mind, your machine shop should be involved when attempting to install thread repair kits on any highly-stressed threads that will bear a lot of load, like the head and main bolt holes in your block.

CHAPTER 3

ENGINE DISASSEMBLY

Now that you've got a place to work in and the right tools in hand, you're probably anxious to get your LS engine apart. To the untrained eye, the teardown of a worn or broken engine might seem nothing more than a necessary evil—the greasy prelude to an engine rebuild, and one that you'd just as soon be done with. But the fact of the matter is, engine disassembly is a critical stage of a successful rebuild project, and this is true in several ways.

First off, hastily tearing an engine down can actually damage parts and render them impossible to reuse. For example, tough-to-turn fasteners may be stuck and cause frustration; but trying to take them out the wrong way can leave you with damaged parts. The result will be time wasted in trying to correct the problem—not to mention the possibility of costly repair or replacement invoices.

Second, failing to take careful inventory of all components as they come out of the engine can make it much harder to diagnose past engine problems and therefore make addressing them far more difficult. For instance, a bent pushrod could indicate rocker arm, lifter, or even valve damage. If you know exactly which parts were working with that pushrod, you'll know what to inspect most thoroughly for possible replacement.

Finally, and especially if you're a novice (and haven't seen many engine components in your day, much less ever rebuilt an engine), stopping and carefully taking a look at all of the parts as they come apart can be invaluable in helping you learn how an engine operates. It will also help you understand where and how parts will fit back together during engine assembly.

We hope the remainder of this chapter helps you understand that if you don't take the time to perform your engine disassembly correctly, you are doing yourself a major disservice!

Organization Techniques

We discussed the tools of the trade and the general techniques necessary for engine rebuilding in Chapter 2, and many of these apply to the disassembly portion of the project. But there are a few areas we need to elaborate on regarding teardown procedures in particular. At the risk of turning this section into a Kindergarten-esque schoolhouse lesson, let's keep in mind the 3 Cs of engine disassembly: cleanliness, cataloging, and care!

Cleanliness

As with any task involving machines, organization starts with cleanliness. Engine disassembly will be by far the dirtiest part of any rebuild project, as used engine parts are invariably covered with oil, coolant, carbon buildup, and other sorts of grease and grime. Keeping things clean will keep you organized, focused, and safe.

Before beginning the engine disassembly process, you should already have chosen that corner of your garage you're dedicating to your rebuild project (discussed in Chapter 2). For disassembly purposes, you'll need to have some sturdy work tables with adequate newspaper or other clean material to soak up oil and other fluids. Have a bunch of paper towels ready at hand, too.

Cataloging

Just as important as keeping things clean is having a reliable parts cataloging system. Each engine part provides a clue that can help you synthesize an overall story of the life of the engine, and this tale is much more difficult to translate if you don't know exactly where each component was removed from. This is even more crucial for parts that are capable of refurbishment and reuse in your new

Engine Removal: Tips and Tricks

With many vehicles, it's possible to strip down an LS engine a substantial amount while it's still installed. The ability to do this can be a boon for maintenance and many high-performance parts installations, but when it comes to a full engine rebuild, the best thing to do is to get the engine out while completely intact–or as close to it as possible. The main reason to do things this way is simple organization. Getting the engine out and away from the vehicle before disassembling it will compartmentalize the tasks; you'll stay more focused and the overall job will seem less daunting. Not to mention, you'll avoid being perplexed by problems like sorting out which bolts were used to fasten unrelated underhood items from which bolts actually were a part of the engine.

Techniques for engine removal differ by vehicle, and because of the huge number of variations that exist for all the GM cars, trucks, and SUVs that were equipped with LS engines, we simply cannot cover them in this book. The best thing to do is have a look in your official GM service manual for the full removal procedure. A few key things to keep in mind, though. One piece of the puzzle that can be most confusing is the wiring harness; the harness may come out with the engine, or many of its plugs may need to be disconnected first. Either way, make sure you meticulously label and take photos of all wire routing–there's always a lot of it, and you don't want to be left with a spaghetti mess that you have no idea how to attach to your freshly rebuilt engine later! Also, exercise extreme caution during the engine removal process; engines (and vehicles) are large and heavy, and you will have to deal with volatile fluids that can harm you or cause you to slip if not dealt with properly. These will always include coolant and fuel, and since some vehicles require disconnecting of power steering and brake lines during engine removal, you might have these fluids around, too. Make sure you collect all such fluids individually (as appropriate) and take them to your local recycling or hazardous waste disposal center.

It can be a messy process, but unfortunately, there's no way to perform a rebuild without removing your engine from the vehicle!

Some engine removal procedures involve the traditional "cherry picker" hoist, while others include dropping the front subframe out of the car (best performed using a lift, though you can usually get by without one). In the latter case, the engine may come out with all of its accessories–and even the transmission–attached to it, sparing a lot of busted knuckles in the tight confines of an engine bay. When reading through GM service manual engine removal procedures, keep in mind that hobbyists have figured out some shortcuts to save time and hassle (see the A/C compressor hanging in the photo? This negates the need to have refrigerant professionally collected and then replaced). You should have a look at How to Build High-Performance Chevy LS1/LS6 V-8s *by Will Handzel for detailed engine removal procedures for 1997-2002 F-cars, 1997-2004 Corvettes, and full-size trucks.*

We strongly suggest that you take photographs of your engine prior to and during removal so that you can refer to them later on during reinstallation. For example, having a visual aid to show exactly how your wiring harness must be routed can be extremely helpful and prevent a lot of head-scratching when it comes time to put everything back together. The diagrams in a service manual are of little help for something like this!

It's an excellent idea to take note of critical bolt locations, wiring hookup points, and other easy-to-forget items as you are going through the process of removing the engine. Here, we can see that the rear of this LS cylinder head was labeled with critical ground wire location and other information as items were disconnected from it, and the bolt for the AIR pump bracket has even been temporarily left in to avoid mixup or loss. Any such markings would need to be photographed for a more permanent record, and any accessory bolts removed and individually tagged, before sending parts to a machine shop.

engine. We'll talk more about all of this in Chapters 4 and 5, but just as an example, finding a rod bearing failure during disassembly would indicate the need to label which rod experienced the problem so that it can be most scrupulously inspected for damage. Photographing of critical steps is also not a bad idea during engine disassembly—we're of course covering all Gen III and IV engines in this book, but sometimes, there's nothing like taking a look at the real thing (for example, your particular engine's VVT system layout) to help refresh your memory.

Small components should be bagged and tagged, but many engine parts are large and bulky (and don't require labeling since it's obvious what they are). Don't just leave the larger stuff lying around, though–generic bins can be helpful to place items in as the engine is being disassembled.

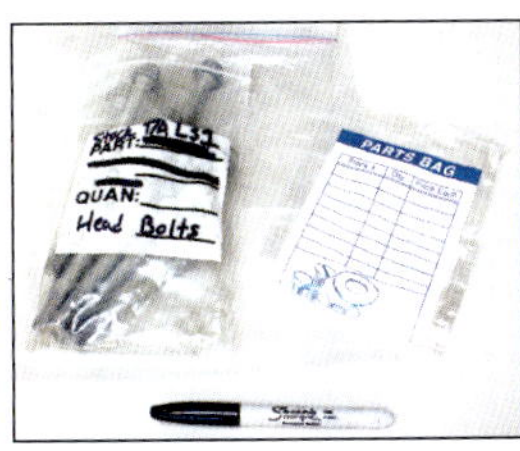

It's possible to purchase various parts organizing trays and other specialized equipment–if you're a total neat-freak (or are in the business of rebuilding engines) this is certainly the way to go. But for others, more rudimentary household items like clear plastic bags and permanent markers (along with other stuff like masking tape and mom's old tupperware) can perform the task of keeping things labeled and organized just fine.

Care

The final piece of the organization puzzle is use of caution, patience, and forethought. For example, a crankshaft is heavy and, if dropped, can very easily break your foot (or if you somehow avert harm to an appendage, your crank will still be costly to repair or replace)! Even for parts you're not planning on reusing, mishandling can cause damage that you might mistake for something caused while the engine was running. And as we talked about in Chapter 2, avoid distractions, it will help you stay safe and organized. You don't want inattention to result in misplacing a critical fastener that you'll need to reuse later.

Step-By-Step Engine Disassembly

Let's go through the complete disassembly of a typical LS engine, specifically an LS6. Because the LS6 is a member of the Gen III family, some of its features and sensor locations are slightly different from Gen IVs. We'll look more closely at generational and other technical variations in later chapters; for now, just know that your engine will probably not look exactly the same as this one (though again, all major LS engine parts—and many minor ones—are very similar). Also noteworthy is that this engine has been equipped with aftermarket exhaust headers, but is otherwise nearly stock.

1 Ready to Begin Disassembly

This engine was installed in a 2004 Corvette Z06, and while the wiring harness and many accessories have already been taken off as part of the process of its removal, others are intact. Still, we've attempted to leave on everything that is best considered a part of the engine itself, thereby giving a typical disassembly starting point for most readers. Also, the oil has already been drained, and you should do the same at this time. Get your Work-A-Long sheet ready, and let's begin!

Documentation Required

2 Disconnect Intake Manifold Accoutrements

The intake manifold is the best place to start the disassembly. Remove all PCV hoses, the MAP sensor, and any remaining wiring from the manifold, carefully marking them with masking tape for later identification. This way, you can put things back exactly where they came from later on. All of these items about the intake manifold vary by vehicle and exact engine type, so take detailed notes and photos!

3 Remove Fuel Rails

On many Gen III and IV engines, it's possible to leave the throttle body and fuel rails temporarily attached to the manifold and remove the entire assembly. Here we opt to remove only the fuel rails for the moment (exact style of fuel rails varies by engine). Four bolts hold them down, after which the rails can be gently pried from the manifold and lifted off. Keep an eye out for details like a fuel rail ground strap on many Corvettes and a fuel rail stop bracket at the rear of the driver side rail on Camaros and Firebirds–it will be tough to figure out where small items like these reinstall if you haven't labeled them.

4 Remove Intake Manifold

Ten 8mm bolts hold the intake to the cylinder heads; remove them and the assembly will lift right up. If the manifold is stuck, a little fidgeting will pull the factory seals from the head surfaces. Thanks to the air gap that exists between the bottom of the intake manifold and the rest of the engine, it's likely that buildup of leaves, oily grime, and other dirt will now be revealed. Be ready with some paper towels, or even a vacuum.

5 Remove Oil Pressure Sensor

The oil pressure sensor is located at the top rear of the engine. In some engines the sensor threads directly into the block (as here), while in others it threads into the valley cover. A special socket tool can be purchased to remove/install this item (even a deep socket isn't long enough for taller style sensors), but an open end wrench can also work just as well if you're careful. Save this sensor for possible later reuse.

6 Remove Knock Sensors

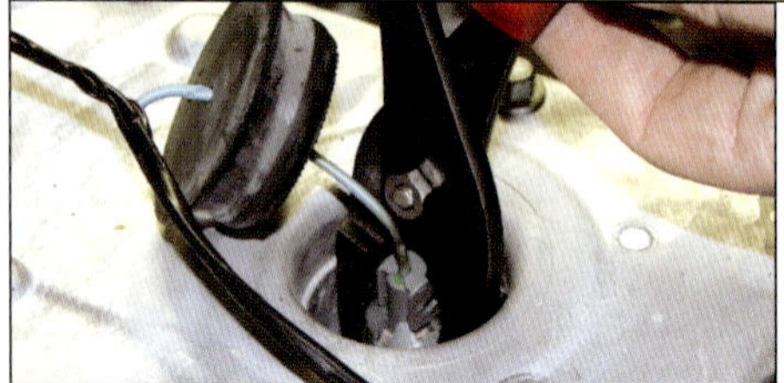

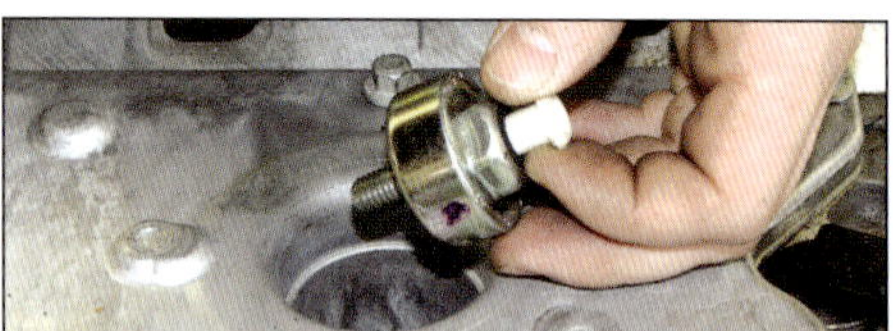

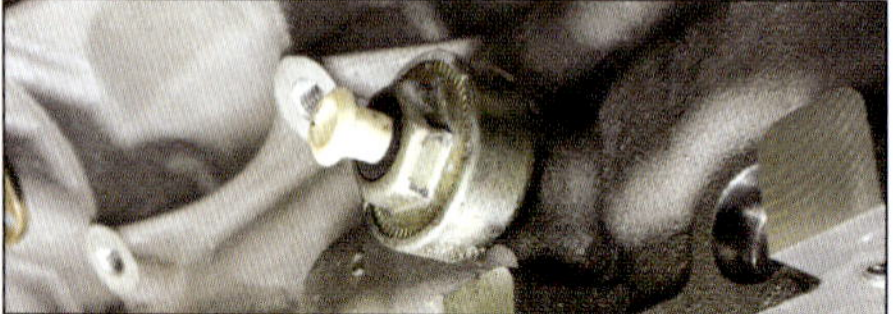

Gen III engines have their knock sensors mounted beneath the intake manifold, in the valley cover area. Pop off the knock sensor wiring harness, but use care on the plastic connectors (pliers gently squeezed and jiggled work well) as they can become brittle in this hot environment over time. Then, use a deep socket to remove the knock sensors. Note that Gen IV engines have their knock sensors mounted down low on either side of the block, as shown in the final photo (note that the sensor shown is actually a Gen III unit installed into the Gen IV location—that block was swapped into a vehicle that was originally Gen-III equipped. See "Block-Swapping Points of Interest" on page 46).

7 Remove Valley Cover

The valley cover can be removed next via ten or eleven 10mm bolts. Some engines have a PCV system integrated into this cover, as does this LS6 unit. Also, some Gen IV engines will have the Active Fuel Management system located here, but because everything in this system is integral to the valley cover (technically called a LOMA on those engines), its removal does not require any additional steps from those shown here.

8 Remove Coolant Air Bleed Pipes

The coolant air bleed (a.k.a. coolant vapor) pipes, which run between ports at the front and rear of either cylinder head, vary in style by vehicle. Some engines have a pipe connecting the fronts of the cylinder heads together and a separate pipe connecting the rears. Others have a single pipe that connects all four (which will require these pipes to be removed before the valley cover), and some don't connect the rears at all, as is the case here. No matter what style you have, they are removed with a 10mm socket and, along with their bolts, set aside for later reuse. You should make a note to buy the correct type of new seal for these (some have an O-ring, while others use plate-style gaskets).

9 Remove Ignition Coils/Brackets

If you haven't done so already, the ignition coils can come off. The individual coils (as well as their wiring harness) can stay on the brackets (if equipped), and each bracket removed as a unit. In this case only remove the bolts holding the bracket to the valve cover; on most engines, this translates to five 10mm bolts each.

10 Remove Valve Covers

All LS engines except 1997 and 1998 LS1s had four centralized bolts holding each valve cover to the cylinder head. (These early LS1s had their bolts around the perimeter.) With the bolts removed, the covers should come off with nary so much as a gentle pry.

Professional Mechanic Tip

11 Remove Rocker Arms

Loosen all sixteen rocker arms via an 8mm bolt on each. Some bolts will turn out more easily than others, as some rockers will be holding their respective valves open at this point, while others will not. It's best to turn each bolt all the way out but leave it in its respective hole. This way, all the rockers on each head can be lifted off while still sitting on the rocker arm pivot supports (shown; note that certain engines such as the LS7 have no such separate supports–their pivot supports are machined directly into the head). Each rocker and its bolt can then be lifted off of the support, labeled, and set aside for possible reuse. Please note: *when removing rockers on an engine equipped with strong aftermarket valvesprings, it is recommended that you only remove each rocker bolt when its respective valve is closed. This prevents possible thread damage in the head.*

12 Remove Pushrods

The sixteen pushrods simply slide up and out of the cylinder head. Use an organizer tray (one of these works well for rocker arms too) or just label each carefully to show where it came from. Reuse is theoretically possible, though LS pushrods are often found to be bent and must be replaced. See Chapter 4 for more information.

13 Remove Dipstick Tube

Remove the dipstick tube (if equipped), which is held to the passenger side cylinder head by a 15mm bolt. Simply pull upward to remove it from its hole in the engine block behind the exhaust manifold. Inspect its O-ring seal for any damage; if there is none, it will be OK to reinstall it on your rebuilt engine.

14 Remove Spark Plugs and Exhaust Manifolds

Remove the spark plugs, then the exhaust manifolds via six bolts a side (if you haven't done so already). Again, this engine was equipped with aftermarket headers, so yours will almost certainly look different. Mark any air injection system tubing and brackets for later reinstallation–not all vehicles had such a system, though this one did, as you can tell from the 2-bolt flange sticking out at the top middle of the header.

15 Remove Transmission Bellhousing / Clutch / Torque Converter

Though we've been dismantling this LS6 as it sits on its Corvette subframe, we'll need to get it onto an engine stand soon. If you haven't done so already, remove everything at the rear of the engine down to the flywheel or flexplate, including the transmission/bellhousing and torque converter or clutch (as shown here). Some light prying may be needed on some of these components in order to facilitate removal, and a pry bar held against one of the engine-to-transmission dowel pins can be used to help keep the flywheel or flexplate from turning while loosening clutch or torque converter bolts. A flywheel holding tool is a better option.

Special Tool

16 Remove Clutch Pilot Bearing

All manual-transmission vehicles had a pilot bearing or bushing pressed into the back of the crank for the transmission input shaft to ride in. Before the engine stand gets in the way, now is a good time to remove it. Different types of puller tools are available on the market, one being this slide hammer—but we advise you save some money and have your machine shop remove and replace this bearing for you.

17 Remove Engine Accessories

Also in the interest of making the engine less bulky and better able to be mounted on an engine stand, remove any remaining large and easy-to-access engine accessories that you may not have already. These might include the water pump, alternator, and power steering pump and reservoir, among other items (depending on application).

Safety Step

18 Mount Engine on Engine Stand

Discussed in Chapter 2, engine stands are readily available at nearly any large auto parts outlet, and now is the time to put it to use (if you haven't already). Make sure you are using a quality stand, hardware, and lifting equipment, and get at least one other person to help you. Bolt the mounting portion of the stand to the rear of the engine block using high-grade, long metric bolts. Adjust so that the engine will be mounted as low in the stand as possible (to best enable later access to bolts at the rear of the engine). Then, using your engine hoist or similar device, lift the engine (don't forget to disconnect the engine mounts!) and carefully slide the mounting portion into the remainder of the engine stand. The engine can now be easily worked on from any angle!

19 Remove All Remaining Engine Accessories and Brackets

With the engine now on a stand, any remaining brackets and accessories that may have been mounted low and difficult to access before can now be removed. These include the starter motor, engine mount brackets, and anything else you may not have removed from the outside of the block already (all of which vary by vehicle).

Notation Required

20 Remove Cylinder Head Plugs and Sensors

Before removing the cylinder heads, take out any screw-in plugs and sensors. This will help avoid damage to these parts once the heads have been removed. On this LS6, the passenger side head has a plug at the rear that is removed with an 8mm Allen key, and the driver side head has a coolant temperature sensor at the front that is removed via a deep socket. Cylinder head plug/sensor type and location may vary, so note where each of these will reinstall later (as well as any ground wire and accessory bolts located at the front and rear of each head, some of which you may have dealt with already).

21 Remove Cylinder Head Bolts

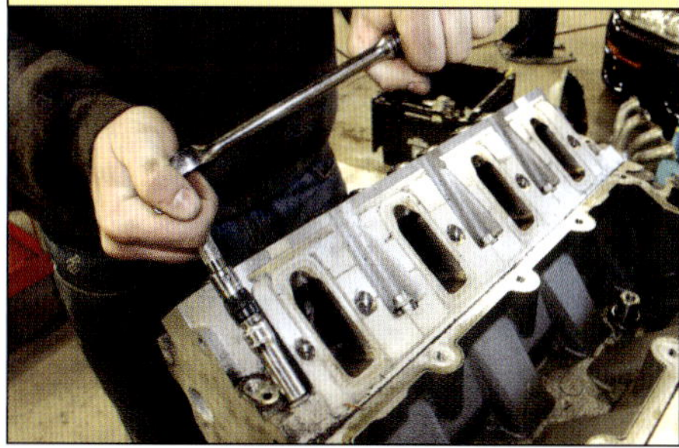

Remove the five M8 thread cylinder head bolts located along the top of each head, followed by the ten larger M11 (or on some engines, M12) thread bolts on each. The M11 bolts will be very tight and will require a breaker bar to dislodge. These M11 bolts can't be reused as they are a torque-to-yield type, but hang onto them for now. You will need to inspect them and determine which type to buy (GM used different head bolt lengths that varied by model year) and you will also need them during pre-assembly. A magnet may be needed to lift the under-the-valve-cover M11 bolts out of their holes as they sit in deep recesses in the head.

22 Remove Cylinder Heads

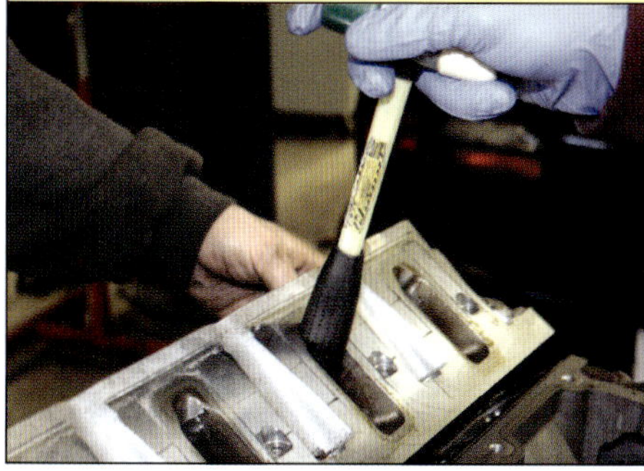

Once unbolted, many Gen III/IV cylinder heads come off easily, but this could also be a slight challenge if the head gaskets have adhered to the block and head surfaces (particularly true of earlier-style, graphite-layered gaskets). Taps from different directions with a rubber hammer can help dislodge the heads. Another technique is to use a hammer shaft in an intake port; just pull it toward you and it should lift the head off. Label which head was which, and carefully set them aside on a soft surface (such as wood blocks) so as not to scratch or otherwise damage their deck surfaces.

Faster Removal Techniques

We talked in Chapter 2 about some of the problems you're likely to encounter with fasteners during a rebuild project (see our Workbench Tip, "Fasteners and Threads: Problems and Solutions"), including some of the issues that you might run into during engine disassembly. We discussed what to do if you find engine bolts that are stuck in place, and also what to do if they break.

However, we wanted to draw specific attention to something many readers may not think of. Many do-it-yourselfers with well-equipped home garages have air-powered impact guns, tools that are very good at removing all sorts of fasteners. You may be tempted to use one of these on some of the tough-to-turn bolts that exist throughout an LS engine, and a good example is the head bolts. However, doing this could spell the end of your engine block, especially if it's made out of aluminum. Use of such a tool can easily damage the threads of the head bolt holes, causing damage that might be too costly to repair or that may not be fixable altogether.

Here's our advice: if a fastener threads into aluminum, you definitely do not want to use an impact gun to remove it. If a fastener threads into iron, you almost certainly don't want to use an air tool to remove it, either (thanks to the risk of thread damage, but also simply to the increased chance of bolt breakage). Probably the one and only safe place to use an impact tool is on the crank bolt of the LS, which is sometimes impossible to remove using any other method (and is threaded into the super-strong crankshaft material).

23 Remove Head Gaskets

Assuming the gaskets didn't already come off stuck to the heads, remove the head gaskets from the block surface. They should peel off easily but may leave some residue (later-style MLS gaskets like those shown will leave much less). Don't worry about scraping off the block surface; aluminum blocks in particular can be damaged if this is done incorrectly, and besides, let your machine shop worry about this kind of cleaning!

24 Remove Cylinder Head Locating Pins

Found at the front and rear of each deck surface of the block, a total of four cylinder head locating pins (a.k.a. dowel pins) help position the cylinder heads and head gaskets on the block surface. They can be removed using pliers. While you can theoretically reuse these, you should probably just discard them and get a new set for later.

25 Remove Lifter Guide Trays

Remove the plastic valve lifter guide trays. There are a total of four, each spanning two adjacent cylinders and held by a 10mm bolt. With the bolt removed, lift the guide up and out. Most lifters (a.k.a. tappets) will come out with the guides, so be careful they do not slide out once the tray clears the block (and possibly land on the floor). Exact guide tray shape varies by engine, so yours might not look quite the same as shown.

26 Remove Remaining Valve Lifters

Some lifters may not come out with their trays, be rather may remain stuck in the block due to sludge deposits or abnormal wear. If needed, use pliers to remove them; but a magnet might do the trick just as well. Remember to mark where each lifter is coming from for diagnosis purposes (also, used lifters are a unique match to their bores, though their reuse is not recommended–see Chapter 4 for more information).

Professional Mechanic Tip

27 Remove Crank Bolt

Install a flywheel holding tool onto the flywheel/flexplate if you have one. Absent access to one of these, an assistant can also hold a large pry bar in the teeth on the outside perimeter of the flexplate or flywheel to help keep the crank from turning (necessary during steps 28 and 30 as well). The 24mm crank bolt will be very tight! While a breaker bar is best to use, this is one case where it is usually OK to use an air ratchet to remove this bolt if you must–it should not damage the crank threads. The crank bolt cannot be reused; however, it will help in later reinstallation of the harmonic damper (among other things), so hang onto it.

Special Tool

28 Remove Harmonic Damper

Remove the harmonic damper ("balancer") with a 3-jaw puller. If your puller does not have the correct cone-shaped fitting to rest in the front of the crank snout, you'll need to thread your old crank bolt in and either keep backing it out as the damper moves forward, or acquire another, longer crank bolt (or one with a ground down outside edge) to push against so the damper can move freely off the snout.

Important Notes:

(1) Check and see whether your engine uses a locking washer between the balancer and the crank snout. If so, make a note that you'll need to install one during final assembly.

(2) Most harmonic dampers are not keyed to the crankshaft on Gen III/IV engines–but thanks to variation on the production line, some have balancing weights in their perimeter. During less extensive engine repairs, you would scribe the harmonic damper's location on the crank so that you'd know its orientation for later reinstallation. However, during a full engine rebuild, you'll be rebalancing the rotating assembly anyway, so you don't need to do this. (See Chapter 5 for more information on balancing.)

29 Rotate Engine on Stand

Up to this point, we've been working with the engine mounted right-side-up on its stand. To provide better access to upcoming components, now is the time to flip the engine over. Make sure the oil has been drained before doing this, and because you will have some amount of fluid dripping out of the engine, have a pan ready to catch coolant and residual oil (a specific one made for an engine stand works best). This type of gooey mix will require special disposal–don't mix it with the oil you're taking to the recycling center! (To minimize coolant splash here, consider removing the engine block coolant plugs before turning the engine over; see step 59.)

30 Remove Flywheel or Flexplate

On most engines, six 15mm bolts attach the flywheel or flexplate to the crank. Once they are loose, you must remove your flywheel holding tool (or pry bar!). Some light hammering around the perimeter will help a stuck flywheel or flexplate come off. You may be interested to note that oftentimes, an "extra" hole in the flywheel or flexplate surface lines up with a corresponding hole in the crankshaft (seen just above the middle and index fingers in the photo) on LS engines.

31 Remove Oil Pan Bolts

The oil pan is fastened to the bottom of the engine block with an array of 10mm bolts, the majority of which are short bolts securing the pan to the engine block (there are two securing the pan to the front cover). But in addition, there are two longer bolts at the rear that extend all the way from the bottom surface of the pan into the rear engine cover. Since some Gen III and IV oil pans are a two-piece design (as is the case with this Corvette pan), don't get confused and remove the bolts holding the pan together–they can stay in for now.

32 Remove Oil Pan and Gasket

Some light tapping and prying may be needed to dislodge the pan, but don't go crazy and damage it: it's a stressed member of the engine! Once it's free, begin lifting up, noting that you may have to jiggle back and forth a bit for the pan baffles to clear the pickup tube that's still attached to the engine (dry-sump engines excepted). As you can see, the pan can be removed with the oil filter still attached if desired (older small-blocks had filters that threaded directly to the block). You'll need to remove its baffles and sensor(s) (to allow you to clean the pan out) later, all of which vary by pan type. For now, cover any oil passage openings, then drill any rivets holding the pan gasket to the pan and throw the gasket away.

33 Remove Rear Engine Cover

Twelve 10mm bolts hold the rear engine cover to the block. The cover will be reused, so don't damage it by prying too hard and scratching the underside; this can result in an oil leak.

34 Remove Front Engine Cover

Eight 10mm bolts secure the front cover (a.k.a. timing cover) to the engine block. Note that the Gen IV front cover is slightly different and has the cam sensor mounted in it–such a cover can be removed with the sensor intact, so the procedure is the same. Some engines also have part of the VVT system mounted here, so you should photograph your engine's particular layout for reinstallation later.

Critical Inspection

35 Note Any Signs of Unusual Engine Operation

With the engine covers removed and the oil-soaked internals exposed, one can have the first good look at anything that looks suspicious in terms of engine wear or damage. Check out the excess metal particles located in this engine's oil pickup screen and the large metal chunks that appeared to be piston ring pieces lying about. These told us to stay on our toes, note what we'd seen, and be doubly diligent in making a record of any component wear or breakage irregularities as disassembly proceeded.

36 Remove Oil Pump Pickup Tube

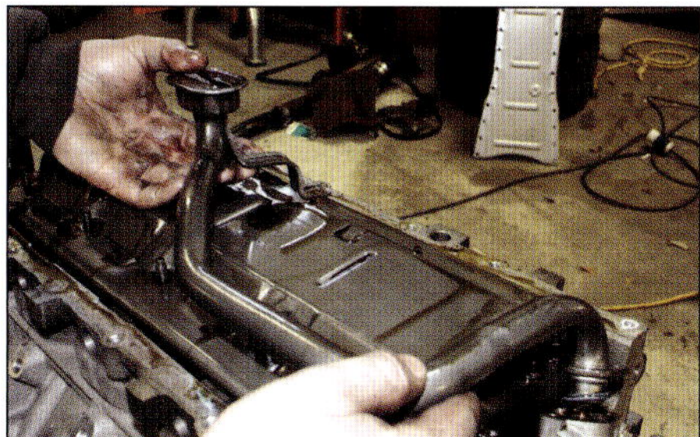

The oil pump pickup tube (GM calls this entire assembly an "oil pump screen") needs to be removed next. A 10mm bolt holds the tube in the oil pump inlet, and a 13mm nut holds the other end of the tube to the crank oil deflector and main cap bolt. You will note that the tube installs into the pump with an O-ring; there were different styles of these used, so hang onto yours so you can compare when buying a new one. (Note that on dry-sump engines, the pickup tube is part of the oil pan, so this step is not necessary.)

37 Remove Crankshaft Oil Deflector

Remove the remaining nuts holding the crankshaft oil deflector (a.k.a. windage tray) to the main cap bolts, then simply lift it off of the engine. This item will be reused, and in Chapter 7, we'll show you how it can be modified if you're installing a longer-stroke crankshaft into your rebuilt engine.

38 Remove Oil Pump

Four bolts come undone and the oil pump slides off the front of the crank snout. We should note that aftermarket bolts and pump spacers have been used here to move the pump forward and accommodate a so-called double-roller timing chain, but most factory setups look very similar. You may now flip the engine right side up on its stand.

39 Remove Timing Chain

The camshaft sprocket is held on by three bolts (or, on later engines, one centrally located bolt). Even with these removed, the sprocket will likely be stuck onto the cam's locating pin. Some prying (don't damage the front cover sealing surface!) and rubber hammer taps about the perimeter will likely be required to coax it off. Once that's done, the cam sprocket and chain slip easily off the crank sprocket.

Important Notes:
(1) Some Gen IV engines also have timing chain tensioners and dampers that must be removed here as well; take note of the layout of any such provisions on your particular engine so that you can reinstall them properly later.
(2) In the case of VVT-equipped engines, the sprocket and attached cam phaser are held on not by a bolt, but by a so-called actuator solenoid valve (finger pointing).
If you will be reusing the GM cam phaser system, take a note to purchase a new actuator solenoid valve, as reuse is not recommended.

Federal-Mogul

Though you're only at the stage of disassembling your LS engine, we thought we'd take a moment to fill you in on a company that can be a source of many parts to use in your rebuild project.

Federal-Mogul Corporation markets products under a few different brand names that are probably familiar to many readers, such as Fel-Pro, Champion, and Sealed Power. As an OE supplier, Federal-Mogul is an excellent source to count on for stock LS rebuilds, but the story doesn't end there, because the company's Speed-Pro brand includes a full range of parts intended more for the high-performance end of the spectrum.

Federal-Mogul is also a great place to look for some engine-related supplies; as we briefly discussed back in Chapter 2, Plastigage is a must-have for your rebuild project, and you'll see it in action during pre-assembly. Also, detailed parts selection discussion will come in Chapter 4, and you'll see several other parts sold under Federal-Mogul brands highlighted there as well! For more information on the company, visit www.federal-mogul.com.

The vast array of engine-related products offered by the various Federal-Mogul brands is more than impressive, it's downright astounding! Bearings, pistons, ring kits, pushrods, and even replacement valves are just a few of the items sold under the Sealed Power and Speed-Pro brands.

During disassembly, you should be making notes of what gaskets and seals in your engine you'll need to replace, and Federal-Mogul's Fel-Pro division is an excellent source for these. Just a couple of examples are the throttle body seal and upper and lower fuel injector O-rings.

Special Tool

40 Remove Crankshaft Sprocket

The crank sprocket is a press-fit design, and chances are you'll need a 3-jaw puller to get it off. As when pulling the balancer, an alternate crank bolt may come in handy to give the puller a surface to push against. You can let the shop servicing your crank worry about removing this sprocket if you like, but either way, the inevitable wear of the gear teeth means that this sprocket needs to be replaced with a new one. Note that factory and factory-style sprockets also double as an oil pump drive gear, and that underneath it the crank is keyed. This key may be stuck in the crank, but if you are able to get it out easily, save it for later reuse or remember to buy a new one.

41 Remove Camshaft Sensor

It's almost time to remove the cam, but before you can do so, a couple of preliminary steps. The first: on a Gen III engine, you must remove the cam sensor at the top rear of the block. Remove the retainer bolt, and the sensor will slide out the top—a gentle screwdriver pry may be needed to dislodge it. (If your engine is a Gen IV, the cam sensor came out with the front cover.)

42 Remove Camshaft Retainer Plate

Remove the four cam retainer bolts (which may be TORX-head) and retainer plate at the front of the engine block. This plate has an O-ring seal on the back that you'll need to inspect for damage. Set it aside for now.

Important!

44 Inspect for Cylinder Ridge

It's time to remove the rotating assembly, but before doing so, take a moment to feel the tops of the bores. On high-mileage engines, sometimes there will be a ridge from the rings wearing along the cylinder walls. If this ridge is exceptionally bad, you can buy a special tool to cut it down, as it may hinder piston removal (the rings will hang up). However, you can also damage the block and render it unusable if you don't use the tool correctly, so the alternative is to simply force the pistons out. This might result in damaged rings or pistons, but you won't be reusing them anyway (and there is almost zero chance the block will be hurt). The choice is yours, make it wisely.

Save Money

45 Turn Engine to Bottom Dead Center

Flip the engine upside-down. A crank turning tool can make the process of turning the crankshaft much easier (particularly later during pre-assembly and final assembly), but if you haven't invested in one at the moment, you can also just install your old crank bolt and use a 24mm socket to turn it (shown). We're about to remove each piston/rod assembly one cylinder at a time, and having a piston at BDC gives the most room to work—so pick a piston to start with and turn the crank until that piston is at the bottom of its bore.

Professional Mechanic Tip

43 Remove Camshaft

As with any pushrod V-8, the camshaft is bulky and long, and there is virtually nothing of it to grab onto for the first few inches as it comes out of the block. To remedy this problem, you can install long bolt(s) into the front so that you have some leverage. Or, as here, insert a long 3/8 extension into the hollow bore in the center of the camshaft (this may not be possible with all cam designs). Keep upward pressure on the cam so that the lobes do not fall onto the cam bearings or other metal parts, possibly damaging the cam (they can touch, just avoid too much jostling). For the first couple of inches, you can also reach around the back of the engine and guide the rear of the cam. Slide the cam toward you slowly until it's out; all cam bearings are the same size I.D. so you'll have to hold it centered the whole way.

46 Remove Connecting Rod Cap

Now, break the rod bolts free with a long-handle wrench or breaker bar. Turn each bolt out a few threads, then stop—don't remove them all the way yet. Tap each bolt head with a hammer to help separate the rod from the cap. This is needed because the rod and its cap will be stuck together pretty well. Then, with the rod bolts still in, begin to hold the rod from underneath. Remove the bolts the rest of the way and set the rod cap aside (along with the bearing shell, though this may stick to the crank journal). Remember not to let go of the rod at any point once the cap is off—though normally the rings will provide enough drag to prevent the piston/rod assembly from sliding out on its own, you can never be too careful!

47 Remove Piston / Rod Assembly

Probably the simplest technique for piston/rod removal involves using the butt end of a hammer against the bottom of the rod. Make sure it's a soft-handle hammer to prevent scratching the crank journals or other components! Hold pressure against one of the flat mating surfaces on the end of the rod and slowly push the piston/rod assembly downward. If the piston doesn't want to move easily, it may need some tapping, but make sure the rod isn't hung up on the block before attempting this (Note: If there is the aforementioned ridge at the top of the bore, this may cause a hang up and require a hard whack to get it out–do not *hit hard if there is anything else interfering.) You don't want the piston/rod to hit the floor as it comes out, so be ready to grab it from below and pull it out the top of the bore.*

48 Replace Connecting Rod Cap

With the piston/rod assembly out, immediately reinstall the rod cap in its correct orientation (this will be fairly obvious for most rods) so that you can't mix it up with the other rods. The same should be done with both bearing halves for inspection purposes. Repeat steps 45 through 48 for each cylinder. Note that most Gen III engines have rods pressed to the pistons, whereas all Gen IVs have floating pins held in by clips; in either case, you can leave the pistons assembled to the rods for now and disassemble them later (or have the machine shop do it for you, which is required in the case of pressed pins).

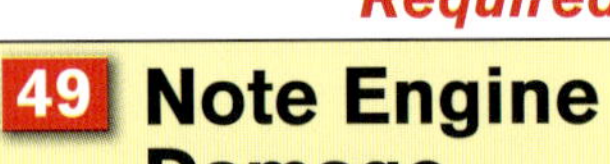

49 Note Engine Damage

One of the ongoing themes of proper engine disassembly is that you should make note of any evidence of wear and tear that looks suspicious; this way, you know what to inspect in more detail later. With this particular engine, however, the root of our previously found particle matter became obvious when the number seven piston was removed. Damage to this piston is not all that uncommon, especially in LS engines that have been exposed to aftermarket forced induction. The consensus is that it has to do with cylinder seven drawing its intake manifold air furthest from the throttle body, and fluid mechanics dictates that it gets a bit more air pressure than the other cylinders (and so runs a little leaner). Add a too-lean custom engine tune into the mix, and you've got a recipe for disaster–and the need for a rebuild! This engine block may have too much cylinder wall damage to be reused, but the machine shop will have to make that call.

50 Remove Crankshaft Position Sensor

It's about time for removal of the crankshaft. First, remove the crankshaft position sensor at the rear of the passenger side of the block. Remove its retaining bolt and gently pry outward as needed. Inspect its O-ring seal for possible reuse.

Important!

51 Inspect for Main Bearing Cap Markings

There are five main bearing caps on the Gen III/IV, and they normally are identified with a factory stamping: number 1 at front, number 5 at rear. These pieces are uniquely machined with the engine block and must not be mixed up, so if there is any question as to the factory markings, make your own markings now using a non-intrusive method.

52 Remove Main Bearing Cap Side Bolts

Two M8 bolts (10mm head) hold each main cap from either side, so there are a total of ten to remove. GM says to discard these bolts; however, they are easily reused if desired and will just need some sealant during reassembly, so bag and label them if you plan on doing so.

53 Remove Remaining Main Bearing Cap Bolts

Remove the twenty M10 main cap bolts, some of which are studded to enable connection of the aforementioned crank oil deflector and oil pump pickup tube. For the outer ones (the ones with studs sticking out the top), you'll need a 15mm deep socket. The inner ones have 13mm heads.

Professional Mechanic Tip

54 Remove Main Bearing Caps

To remove the main caps, GM sells a special tool that utilizes a slide hammer. But we recommend simply using two medium-size 3/8-inch extensions or pry bars to lift each cap gently from the underside. A back-and-forth rocking motion can also help, so have the extensions or pry bars facing opposite directions if need be (not shown). The caps will be tight in there (thanks mostly to the matched machined fit between each cap and the block), and this is particularly true of the thrust bearing cap (number 3, shown being lifted out). Remove the lower main bearing shells as well, which may stick to the crank journals, and keep them with each respective cap for inspection purposes.

Safety Step

55 Remove Crankshaft

With the main caps removed, the crankshaft is free to come out. Use of gloves with a grippy surface is strongly recommended thanks to the heavy nature of the crank, but at least clean the surfaces you'll be grabbing to get the oil off. Lift straight upward and wiggle the crank a little to get it out. Don't let any of the rod or main journals get scratched. And don't drop it! Set it aside in a safe place, preferably in a dedicated crank box or on wooden V-blocks. Note the reluctor ring at the rear of the crank: we'll talk more on options regarding this item in later chapters, but for now, just be careful not to let it contact anything as it can easily be damaged.

56 Remove Upper Main Bearing Shells

Remove the upper main bearing shells that are still sitting in the block, using a screwdriver to carefully pry them out as needed. Label where each one came from for diagnosis purposes (keeping them with that crank journal's main cap is best). They will not be reused, however.

57 Remove Metal Engine Block Oil Gallery Plugs

The block is now nearly bare, and you can now remove all remaining engine block plugs, starting with the metal oil gallery plugs. Some of these plugs have washers that, if undamaged, can be reused, so don't lose them if for some reason they are not attached to the plugs. Exact location of plugs may vary by engine, but the ones you'll typically see are: Front oil gallery plug (press-in, far right of photo): This must be removed to enable full cleaning of oil passages. As a long rod is required to hit it from the back, you may leave this one to your machine shop. Front left (driver side) oil gallery plug (shown being removed). Rear left (driver side) oil gallery plug (not shown here; location is near upper right corner of photos of next step).

58 Remove Rear Oil Gallery Plug

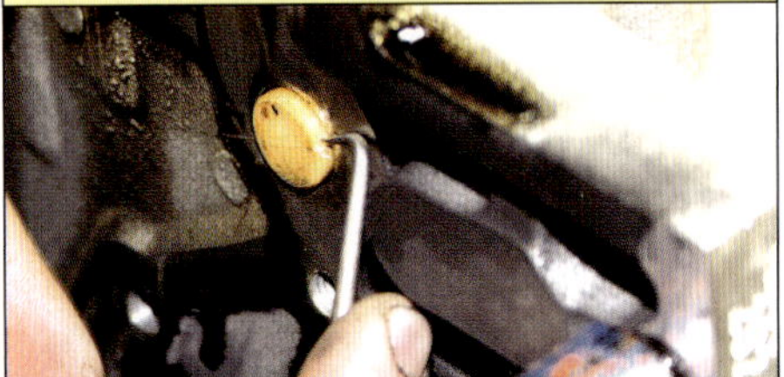

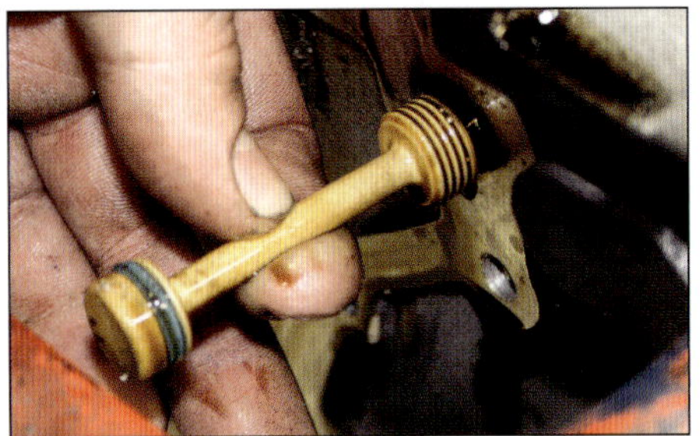

The rear oil gallery plug (which is made of plastic and sometimes called the barbell restrictor) gets sandwiched into its place in the block behind the rear cover. A pick or similar instrument will likely be needed to coax it out of its hole in the main oil gallery. If its O-ring is OK, it can be reused; but this plug is inexpensive and should be replaced.

59 Remove Engine Block Coolant Plugs

The final block plugs to remove are the coolant plugs, which include: Right rear (passenger side) coolant plug (top photo, small plug). Driver side coolant plug (bottom photo, large plug). This plug will be located toward the rear of the block on some engines, and toward the front on others (shown). It is also the location of the engine block coolant heater on vehicles so equipped.

60 Engine Disassembly Complete

With only the cam bearings still in place in the block (leave their removal to your machine shop), we're down to the bare block, and the first important step in your rebuild project is complete (except for cylinder head disassembly). In Chapter 4, we'll take a closer look at the parts we've just removed and talk about options for repair, replacement, and upgrades.

Cylinder Head Disassembly

We removed and set aside the cylinder heads earlier in this chapter, and because these components will require machine work in order to freshen them (if they're being reused at all), you may wish to leave the task of their disassembly to a machine shop; high-performance operations like porting and polishing aside, you'll at least need a shop to clean your heads up, machine the valves and valve seats, and perform other processes that we'll go over in Chapter 5. But if you are dying to remove your valvesprings and pull your valves out at home, here's how!

1 Ready to Begin Cylinder Head Disassembly

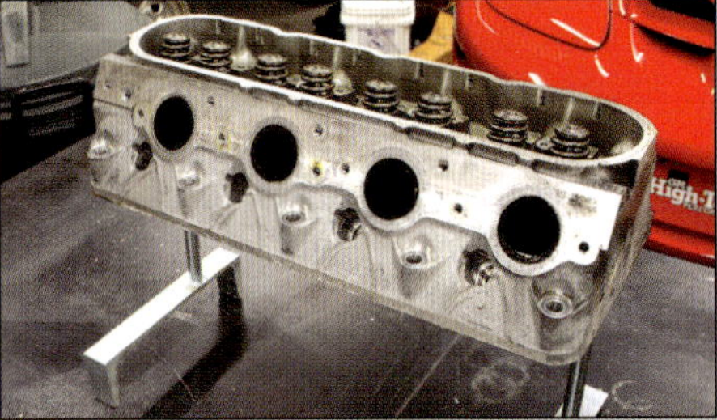

After a quick cleaning with some dirt-dissolving solvent, the head is ready to disassemble. While this process can be performed without them, it's easiest if you use some sturdy cylinder head stands like these from Goodson.

2 Loosen Valvestem Locks

Over time, valvestem locks (sometimes called valvestem keys) can become stuck to the stem or the surrounding retainer. A good way to dislodge them is to put a socket atop the retainer and give it a sharp whack from a hammer. Unless your valvesprings are very weak, this should not cause the valve locks and other parts to go flying everywhere—but be sure you have your safety goggles on just in case!

Cylinder Head Disassembly *CONTINUED*

3 Adjust Valvespring Compressor

Set your valvespring compressor tool roughly in place on the spring you'd like to remove. A good tool (like this one from Goodson) will have adjustable "jaws" at the top, which can be spread open or closed to just clear the valve locks, yet rest firmly atop the retainer. The bottom portion of the tool should have a knob that is turned to allow it to rest against the underside of the valve. Follow the instructions provided with your tool to get yours properly adjusted (so that it locks firmly in place once the valve is compressed). Also make sure you are using a large and sturdy enough tool–the LS is not a lawnmower engine!

4 Compress Valvespring

With the tool adjusted, the valvespring can now be compressed. Make sure the bottom portion of the tool is sitting squarely on the underside of the valve, and squeeze the tool handle. The compressor should lock in place, pushing the valvespring and retainer far enough down to give you access to the valve locks.

5 Remove Valvestem Locks

A small flathead screwdriver will be very helpful in removing the locks from the valvestem (one that's magnetic is even better). Some gentle prying may be required – if a lock is really stuck to the valvestem, you didn't hit the retainer hard enough with the hammer! Also try pulling sideways on the top of the valvespring compressor to afford more clearance between the stem and retainer if necessary. Once the locks are out, release the valvespring compressor tool locking mechanism to take pressure off of the valvespring (be prepared for a little jerk, and keep your safety glasses on despite valvespring flyage being virtually impossible). Although unlikely, be prepared to catch the valve if it falls out the bottom of the head.

6 Remove Valvespring and Inspect Valvestem

The valvespring and valvespring retainer (sometimes called a keeper or cap) can now be lifted off. We can now see the valvestem and valvestem oil seal protruding from the head; engines with a lot of mileage occasionally have metal burrs toward the top of the valvestem. These can theoretically occur as a result of wear between it and the valveguide, but most commonly, they occur as the valve locks beat upward on their grooves in the stem over time. If present, they'll need to be filed down as they'll prevent valve removal (or worse, damage the guide on the way out, possibly making the guide unusable and costing you more money). This one is fine.

Cylinder Head Disassembly *CONTINUED*

7 Remove Valve

Push down on the valvestem and the valve will slide out the bottom of the head. If there is any resistance to movement, don't force the valve out! See the previous step and take action accordingly. Label all valves as they come out, or better yet, use a valvetrain organizer tray like this one from Goodson—it well help you keep track of which valve (and its accompanying parts) is which for diagnosis purposes.

8 Remove Valvestem Oil Seal

Use a valve seal puller like this one from Goodson to grab onto the top of the valvestem seal. A hard clamping action distorts the metal portion of the seal and weakens its grip on the guide, after which the seal can simply be lifted off of the guide and head. Note here that this particular engine used a one-piece seal and valvespring seat assembly—exact design of these components varies by engine and year.

9 Cylinder Head Disassembly Complete

Repeat these steps for all the other valves in both cylinder heads. We've done about all we can with the heads at home, save for some cylinder head inspection that you'll see in Chapter 4.

PARTS INSPECTION AND SELECTION

So your engine is all apart, and you've got containers and boxes of old parts seemingly everywhere. How much of it should you, or can you, reuse? A lot of this has to do with the state of wear your engine parts are in, but it depends even more so on the desired characteristics of your end product. Do you simply want a stock engine with restored stock efficiency and performance, or are you looking to put together a 650-hp high-performance LS? You'll need to decide this well before you start picking through online catalogs, or loading up parts and heading to a machine shop.

This chapter runs the gamut of all the major (and many of the minor) parts that make up a Gen III/IV engine, and is meant to give you as much guidance as possible in selecting the right components for your application. We'll tell you what to look for when you might want to reuse a part, and also rule out items you cannot. In some cases, we'll demonstrate how to use tools we discussed in Chapter 2 to help verify the quality of your parts. While many of these tools are by no means mandatory, they can help save you time and aggravation since you'll have a better idea which parts your machine shop will succeed in refurbishing for you, and thereby avoid phone calls telling you that your parts are no good. We'll also give advice on what parts you might think about substituting in a high-performance engine. In fact, we've scattered information throughout the book (much of it in this very chapter) that focus on parts for making and withstanding high amounts of horsepower!

Parts Selection Words of Wisdom

You probably had a decent idea of what kind of engine you wanted to put together before even beginning to tear your stock LS apart. It's time to chisel that rough vision into a very clear and precise one. High-performance or not, *now is the time to select, and acquire, each and every one of your engine parts*. Choosing the combination of components that's right for you is critical to the success of your rebuild project, and the individual parts need to closely complement one another.

Lack of attention to detail here can not only cause parts mismatch leading to lost power and efficiency, but in severe cases, it can lead to eventual engine failure, or even make your engine impossible to assemble! Imagine spending hundreds or even thousands of dollars on parts and prep, only to discover during pre-assembly that some of your rotating assembly parts simply won't work together. Being stuck with expensive parts you've already had machined (or otherwise have rendered non-returnable) is a nightmare scenario you most assuredly want to avoid.

Be advised, however, that while doing your homework here can assist greatly in avoiding parts compatibility problems, there is always the chance that some additional effort will be required in getting them to fit together. *Minor parts fitment issues are a normal part of any engine build.* The distinction lies in the key term, "minor." For example, it's not unusual for aftermarket rocker arms to require *minor* modifications be made to factory valve covers to obtain the proper operating clearance. Less significant problems like these are often unavoidable, particularly when combining aftermarket parts made by different companies (and we will deal with many of them in Chapter 7, Pre-Assembly). The type of scenario you would want to avoid *right now* is purchasing rocker arms that are not physically compatible with your chosen cylinder heads; for example, buying a standard set for later-style heads that require an offset intake rocker. An inattentive choice like this could result in a not-insubstantial headache later on, so putting the time and effort in *now* will keep the major problems to a bare minimum.

Finally, the blossoming parts market for LS engines continues to expand on a monthly (if not weekly) basis. To keep yourself up-to-date on the latest component availability, it helps to check out LS-oriented websites as well as those of aftermarket parts companies (listed in our source guide). Magazines like *GM High-Tech Performance* will also help keep you tabbed on the latest trends. Do your research, and you can select a combination of parts that will not only work together in your engine, but will yield the power and reliability you've been dreaming of!

Engine Block

The largest part of a Gen III/IV engine is also the one that's most easily and commonly reused. The vast majority of the time, an engine block can be refurbished by a machine shop and made as good as new, or better! This is particularly true if this is the first time your LS engine has been disassembled (that is, the block has never been used in a rebuild before), as this means there should be plenty of metal left on surfaces that need to be re-machined. We'll discuss the entire block machining process in Chapter 5.

Many truck LS engines used iron engine blocks for enhanced durability without concern for weight savings, like this Gen III LQ9. Like many other LS blocks, this is a great engine block to reuse; don't let a little outside surface rust turn you off! (Also, because of their advantages in strength, those looking to build a very high-horsepower LS may be interested in swapping out an aluminum block for an iron one like this. You should be able to pick one up used or even buy one new from GM, more on all this in a moment.)

Don't know whether your engine block has been used in a rebuild before? You can use a dial bore gauge to roughly measure your cylinder bore and see if it's near stock spec (stock bore sizes are listed in the engine RPO table in Chapter 1). Oversize of a few thousandths or more means the block has probably already been overhauled. Since there is a practical limit to how far Gen III/IV engine block cylinder bores can be resized (the exact amount varies by block type and year), this could indicate insufficient material left to work with in the cylinder walls. This is particularly true in the case of aluminum blocks, whose pressed or cast-in iron cylinder liners are often thin.

If you discover (or already know) that the block has been used in a prior rebuild, don't sweat it yet. Depending on how much material was removed from the bores and other areas in that prior rebuild, you may be able to reuse the block again. Your machine shop may have the final say on this, as they can measure surfaces with precision tools and also have the experience to know "how far you can go" with a given block, and the physical limits of what their machining equipment can accomplish.

There are other issues that can render a block unusable. In some cases, block damage may have occurred during operation due to engine part failure or improper maintenance. The potential for physical harm is particularly high for engines that didn't run correctly (or at all) prior to disassembly. Depending on the nature and extent of any damage, the block may be beyond hope. For example, internal engine component failure may have caused part of the block casting to crack or be broken completely. Or perhaps the engine was run completely out of oil, overheating the main bearings to the point where the block's main bearing bores are too far out of shape.

If you and your machine shop determine that your block can't be reused, it's not the end of the world. You can likely find a good used block on the relative cheap if you hunt around. If you find yourself in this situation, try to find a

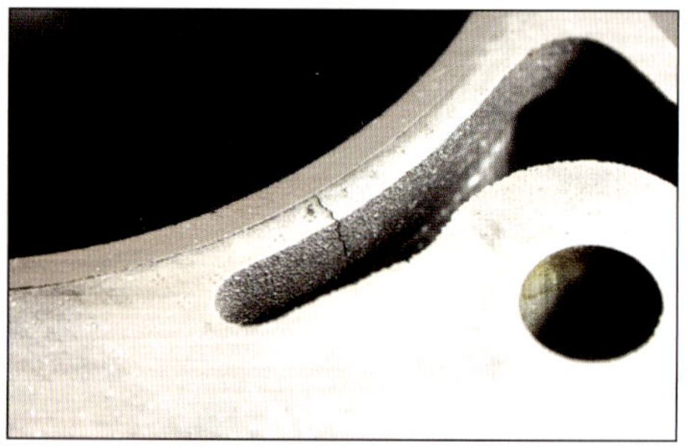

In general, aluminum blocks are more susceptible to cracks than iron blocks, but you should carefully inspect every inch of either block type for them. Top photo: damage at the rear of this LS2 block was actually caused by a motor vehicle accident, but similar cracks can occur by bolt overtightening, internal engine component failure, overheating, or even simply on their own in a normally running engine. Bottom photo: it's not uncommon to find cracks in the areas around the cylinder liners, and many times you won't be able to see them until any head gasket residue is cleaned off of the deck surface (again, this is something you should have your machine shop do).

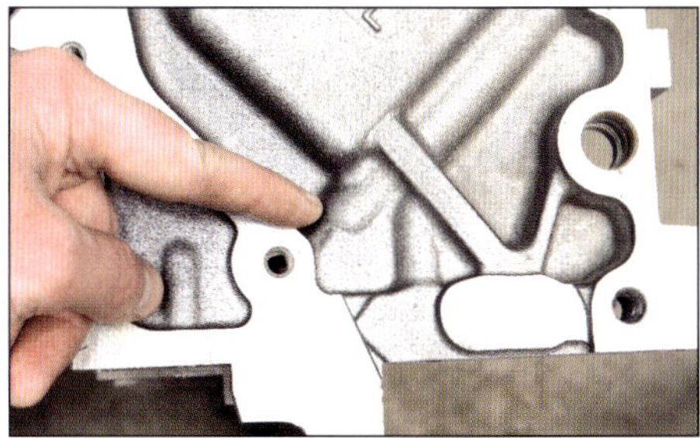

Another area to be concerned about is the bottoms of the head bolt holes. The inner ones are not as easy to see, but the rearmost ones are easily viewed at the back of the block. If an engine's cylinder heads had ever been removed and replaced (during service or high-performance upgrades), chances are that some coolant flowed into the head bolt holes. Failure to properly blow these holes out would cause hydraulic pressure to build when bolts are reinstalled, often resulting in block cracking in this area. Fortunately, this block is fine.

Some amount of cylinder wall marking is normal in an engine that has run for any amount of time. However, cylinders can be damaged beyond repair in certain situations. A broken piston or ring dragging up and down along a cylinder can sometimes cut a groove too deep to enable block reuse. Here, this aluminum-block engine was run low on oil, resulting in loss of splash oiling from the rod bearings. Though wear markings are clearly visible on this bore, they're fortunately shallow enough to be cleaned up with a hone. We'll talk about the possibility of cylinder resleeving in Chapter 5; but the expense and dangers involved in this process mean that it isn't normally feasible or cost-effective.

Other engine problems can lead to some less common block damage. For example, a leaky head gasket can cause hot combustion gases to seep between the block and head surfaces. With an aluminum block, these gases can actually melt the area just outside the iron cylinder liners, as seen here.

block that's as complete as possible. Engine bolts are optional, but at a minimum, you'll need a block that still has its original main caps (numbered 1 through 5). Purchasing and machining new caps to fit can be very expensive and will leave you wishing you just bought a whole new block. Also, one of the most important considerations when shopping for a used block is to make sure to find one that hasn't been sitting exposed to the elements. Dirt and grime will usually wash off, but what you should be most concerned about is moisture. Obviously, iron blocks are most susceptible to corrosion, but parts of an aluminum block can rust too. However, even rust-coated cylinder walls may be OK if it hasn't eaten its way too deep.

GM also sells brand-new, bare LS engine blocks in several different bore, material, and equipment (ex., AFM-ready) configurations. As we mentioned in Chapter 1, a great thing about the Gen III/IV engine family is that there is so much cross-compatibility of parts, and this is a boon for high-performance. Some blocks offer potential advantages like improved bay-to-bay breathing windows (for reduced pumping losses) over earlier LS castings, but this is only the beginning. Want to swap your iron block for an aluminum one and save some weight on the front end of your vehicle? Or perhaps you'd like to substitute your 3.898-inch bore LS1 block for something with a larger bore, like a 4.000-inch LS2 or 4.065-inch L92? No problem—most engine parts will fit (with the appropriate changes like piston size of course)! We'll discuss any parts substitutions you might need to make as we proceed along, but by and large, most engine components will work almost across the board.

And if you're really serious about power, you should know that factory aluminum blocks are generally good for (at a minimum) a couple hundred extra horsepower, while iron ones even more than that. For the ultimate in reliability and power potential, you might even think of stepping up to an aftermarket engine block. Check out our "High-Strength Blocks from GMPP" Workbench Tip for more information.

Rotating Assembly

The term "rotating assembly" generally refers to an engine's combination of crankshaft, pistons, and connecting rods, but it can also include related paraphernalia like piston rings and bearings. All of these items need to be selected in concert, as each must be compatible with the others in physical dimensions, mass, and material. In addition, once all rotating assembly parts are selected, they'll need to be balanced (a process we'll see in Chapter 5).

When selecting each of these components, the money you'll need to spend will rise with the amount of horsepower you're looking to make. As we'll explain below, there are a lot of quality components that GM put into Gen III and IV engines from the factory, and for many stock and moderate rebuilds, you're often just fine reusing some of them. As we discuss each of the items that make up a rotating assembly below, keep in mind that it's easy to go overboard and spend a lot of money on parts tougher than you need for your application. Perhaps just as significantly, it's possible to "cheap out" and buy parts from less reputable companies. There is a difference when it comes to internal engine parts, and it is far better to spend a little extra money on a

Block-Swapping Points of Interest

To a good degree, the cross-compatibility of engine parts between LS blocks also holds in cross-compatibility of electronics. This is particularly true if you're swapping your block for one of the same generation: Gen III for Gen III, or Gen IV for Gen IV.

There are, however, a few things to keep in mind having to do with engine electronics, especially when crossing the generational gap. The age of modern engine control may have brought improved engine efficiency and power, but it can also cause some headache when manufacturers like GM update their electronic systems! Some of these issues are more vehicle-related than they are engine-related, so the exact specifics are beyond the scope of this book–nonetheless, here are some tips to help get you started:

Camshaft position sensor (CMP)

Gen III blocks have the cam sensor at the top rear of the block, while Gen IVs have it in the front cover. Wiring harness extension kits are available to convert one style to the other, and it is also possible to use a Gen IV-style front cover and sensor on a Gen III block. However, you must be sure your cam has the rear-located sensing ring if using a Gen III-style cam sensor. Similarly, your camshaft sprocket will need the appropriate lands and grooves if using a Gen IV-style sensor. See our valvetrain section later in this chapter for more information.

Knock sensors

The Gen III's knock sensors were located in the lifter valley, but for the Gen IV, their placement was changed to down low on either side of the block. The sensors themselves were also revised to compensate for the different engine acoustics at this location. Wiring harness extension kits are also available to convert between these (some kits also use one generation's sensors in the other's location), but ECM or PCM recalibration, modification, or retrofit may also be required.

Crankshaft position sensor (CKP)

Though crank sensor location is similar in all blocks, GM began introduction of more advanced engine computers during model year 2006, and this required a 58X crankshaft reluctor ring as well as an accompanying different crank sensor. Earlier engines had used a 24X ring, and to make things more confusing, the break between 24X and 58X did not correspond exactly with generational switch to Gen IV (early Gen IVs still used the 24X). Fortunately, you can swap between the two ring styles by pressing the ring on and off the rear of the crank (you will see this in Chapter 5). Also, most aftermarket cranks can be had with either style ring, so it is relatively simple to match the required ring up with the needs of a given ECM or PCM.

Active Fuel Management

No Gen III block has the provisions for AFM (a.k.a. DOD), and while most Gen IV blocks have the drilled passages needed to accommodate the system's valley-mounted LOMA, this is not true of all blocks. If you plan on keeping your AFM system (or perhaps even adding the necessary electronics to make your vehicle AFM-capable), you will need to choose a block that can support it.

company with a good name. This way, you can rest easy knowing the parts will stand the test of time—and your lead-weighted right foot!

Crankshaft

One of the most expensive parts of an engine is the crankshaft, and the good news is that all Gen III/IV engines were equipped with excellent cranks from the factory. Even the weakest of LS cranks are cast from a nodular iron material, so they are high quality and popular to reuse. Some engines (like the LS7) even had forged steel cranks for high horsepower with even greater reliability. Though Gen III/IV crankshaft stroke, counterweight design, and material varies between engines, all main and rod bearing journals are a common size across the LS engine family (2.558 and 2.099 inches, respectively), enabling one crank to work within virtually any block and with most any connecting rod. If you're thinking of swapping your crank for a different factory unit, here are a few of the more notable things to keep in mind:

If in good condition, reuse of your factory crankshaft can be an excellent route to take. This doesn't just apply to your standard rebuild, either: thanks to high-strength features like journals with undercut and rolled fillets, many LS engine shops and tuners have found them to be good to 600 hp or more! Use extreme care when storing and/or transporting your crank to your machine shop–a special crank box presents the least chance for damage.

High-Strength Blocks from GMPP

Shop around, and you'll discover more than a few high-strength LS engine blocks available for extreme street and racing applications like large cubic inches and high-boost forced induction. But the best source for high-performance engine blocks, hands down, is GM Performance Parts.

The company's Aluminum C5R Racing Block is the same championship-winning foundation first used in the original C5R race car. This Gen III-style block features siamesed water jackets and special cylinder liners, enabling a max cylinder bore of 4.160 inches. That means this block is capable of over 440 ci! But there's no point in building a large displacement engine if it can't be durable, and that's why the C5R is cast from 356-T6M aluminum, fully x-rayed for defects, and CMM'd to verify accuracy. Best of all, production-size cam journals as well as a production-spec head bolt pattern and cam location mean that you can bolt up just about any production-style parts.

If you're looking for a block that can support even higher horsepower, look no further than the LSX Bowtie Block. This CNC-machined iron block can support up to a 4.25-inch bore, yet has the same 4.400-inch bore spacing shared by all LS blocks. Though it also shares a standard block's head bolt pattern, it comes ready to accommodate an additional two head bolts per cylinder for additional clamping power with extreme cylinder pressures (from an enthusiast's standpoint, the change to the four-per-cylinder head bolt pattern brought by the Gen III has been one of the LS family's few lamented features; previous small-blocks had five). A priority main oiling system ensures total crank protection, and with an external orange powdercoat, the LSX looks cool, too! Plus, a tall-deck version of this block is available, meaning huge crankshaft strokes can be accommodated—enough to support well over 500 cubic inches!

A full listing of these blocks' impressive features could fill pages, so be sure to check out the latest GM Performance Parts catalog for full information. And with more bulletproof engine blocks promised in the years to come, one thing is clear: if you're in search of a basis for the ultimate high-power LS, GMPP has you covered!

The C5R is an absolute work of art, and well worth the money if you're looking for a solid foundation for a big-cubic-inch LS. GMPP says this block is good to at least 900 horsepower, and because it's aluminum, it's also lightweight.

Down at the bottom, C5R blocks include heavy-duty 6-bolt steel main caps for high strength. In addition to sharing the six-bolt configuration of all LS blocks, they are also doweled to help prevent side-to-side cap walk under extreme power and RPM. Premium main cap fasteners are included.

Made from 280 kPa tensile strength cast iron, the LSX has all of the features you'd want from a high-horsepower-capable block—and then some. Aside from its larger cubic inch capability, another advantage the LSX has over the C5R is that it's easier to own. Check with your nearest GMPP supplier for current pricing—we guarantee you'll be surprised at how affordable this block is! (Photo courtesy of GMPP)

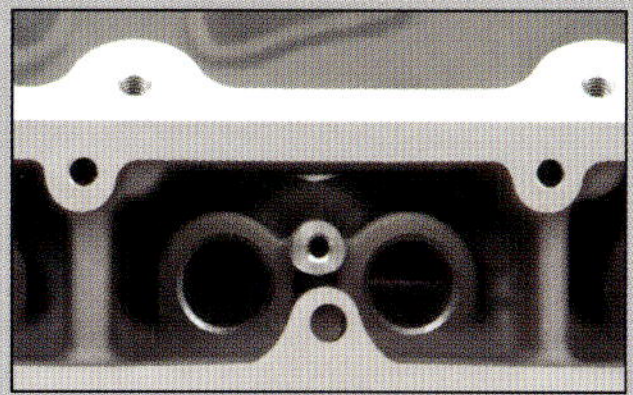

While the additional fifth head bolt for each cylinder threads into cast-in areas below each bore of the LSX, the sixth head bolt created a bit more of a challenge for engineers. The solution: use a stud threaded into the head and secure it against this tab protruding from the upper edge of the block's deck surface. This approach avoids adding stress to the valve lifter area (as would be the case if a threaded hole were simply added down between the lifters). Remember, it is not mandatory to use these extra bolt holes, and any LS head will bolt to the LSX! (Photo courtesy of GMPP)

Rotating Assemblies: Critical Considerations

If you're not reusing any of your stock internal engine components, you've got a good bit of homework to do. Selecting the correct combination of items to comprise a rotating assembly that will work for your application can be quite a task, and lack of careful consideration here opens the door for potential parts mismatch. In addition to being of different strokes, aftermarket crankshafts are often manufactured with varying journal shapes, so this can dictate the range of bearings that can be used. Different rod designs and lengths also affect piston specifications, not to mention crank counterweight requirements. Plus, after juggling rod length with piston pin height, you have to throw all of this into your compression ratio equation (see the Appendix for these respective calculations). And again, even once the items are in your hands, all reciprocating masses must be precisely matched via balancing.

Experienced engine builders can make all of these choices virtually in their sleep, and there's no reason why you can't do your research and come up with a good, complementary set of components yourself. But what may be even better is that you can save yourself some frustration and worry by buying an entire rotating assembly as a kit. These kits take a lot of the potential for error out of the component selection equation, and many times they even come balanced. Not only does foregoing the need to balance save you some money, but many kits also offer up-front discounts compared to buying each of the components individually.

Rotating assembly kits for high-horsepower and budget builds alike are readily available in many different configurations to fit just about any Gen III or IV block. One great source for high-strength rotating assemblies is Lunati. The company offers kits designed to fit the needs of any LS build, with varying crankshaft strokes and piston options to help achieve the displacement and compression ratio the customer demands. The kits also include piston rings as well as main and rod bearings, making them 100 percent complete.

Don't think a rotating assembly kit is right for you? That's OK–read on and we'll give you the best guidance we can on how to individually select the internal engine components that are right for you.

This particular rotating assembly is a Lunati LS1 Pro Series Stroker kit designed to get 383 ci from an LS1 or LS6 engine block. These assemblies are a matched set and include Lunati's own Pro Series 4340 non-twist forged steel crank, Pro Billet Super Light connecting rods (also aircraft-quality 4340 forged steel), and forged aluminum pistons. Rated at a whopping 1,100 hp, its suggested retail price isn't exactly cheap, but the kit is a huge savings over purchasing each of its components individually.

Rear flange width: Most factory cranks use a narrower flange thickness where the rear crank seal rides. However, some early 4.8L and 6.0L Gen III truck engines used a crank with a wider flange. The difference had to do with varying GM transmission dimensions. The short-flange cranks are actually compatible with all transmissions (you can buy a flywheel/flexplate spacer kit from GM if need be), so you should steer clear of wide-flange cranks in order to avoid a potential incompatibility on this front.

One thing to keep in mind if you're pondering the use of a factory crank is that despite their identical dimensions of journal size and spacing, some cranks had a wider rear flange than others. Shown is the narrower, 0.857-inch flange; wider flanges measure 1.250 inches. Also, engines installed in FWD applications (like the LS4) were an additional 3mm shorter in this area, so swapping a non-LS4 crank into an application destined for FWD can be a problem.

Snout length: Make sure the crank that you're looking at has the correct snout length for your application. Some examples of particular concern: LS4 engines had a 10mm shorter crank snout to help further alleviate the packaging concerns inherent with FWD applications. Also, dry-sump engines such as the LS7 and LS9 had longer crank snouts to accommodate those engines' more complex oil pumps.

Reluctor ring type: To aid compatibility with your engine electronics, it makes things easier if you pick up a crank that already has the correct 24X or 58X reluctor ring installed onto the back (even though you may have to replace it anyhow if it's determined to be out of shape). Also, very early 24X reluctor rings may not work properly with all electronics and should be avoided. Aside from the earlier rings being unrecessed, you can tell them apart because the two ring "halves" are very narrowly spaced. (The crank flange photo on the previous page has an early style 24X ring, while the ring comparison photo below has the updated 24X unit.)

Here is the difference between a 24X (right) to 58X (left) reluctor ring, sometimes called the reluctor wheel or exciter ring. These items are reportedly easy to get out of whack, and if they're even slightly warped or otherwise out of shape, your crank sensor won't get a proper signal. Fortunately, they're replaceable, and you can buy either style ring at your GM dealer.

Counterweight design: Even crankshafts of the same stroke used varying-style counterweights to compensate for the pistons and rods used in different engines. Though having a crank with the wrong counterweights probably won't make balancing your rotating assembly impossible, it could make it substantially more expensive. See Chapter 5 for more information.

Drilled vs. undrilled: Many truck cranks do not have the centerline of the crankshaft drilled out. This makes the car cranks more desirable, as they're better at evening out crankcase pressure (and are also a little lighter). This drilled hole does not affect strength.

Regardless of whether you're going to use your own crank or another factory unit, used crankshafts need to be refurbished before they can be incorporated into an engine rebuild project. At a bare minimum, they'll need to be cleaned and have their journals checked and polished, and we'll do all of that at the machine shop in Chapter 5 (along with balancing). But for now, look at the photo captions for some tips on what you can do at home to see if your crank is even capable of reuse. The main items to look for are excessive scratching or scoring on any of

Many crankshafts will exhibit little, if any, visible wear on the journals. However, wear can be significant on cranks with a lot of miles or from an engine with a poor maintenance record. This one is particularly bad off as it even has bearing material embedded in it. This crank could still be salvageable; even though GM does not recommend grinding its cranks, many engine machinists will tell you this is baloney. You'll need to have a machine shop take a look at a crank like this to determine whether it's toast.

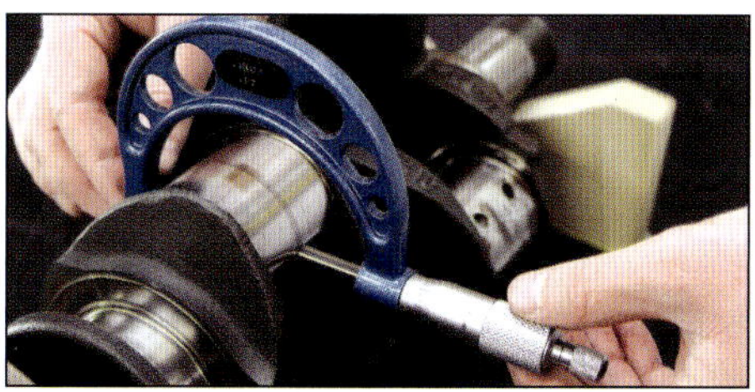

Some cranks will have problems that go beyond what can simply be inspected with the naked eye. Out-of-round or tapered crankshaft journals will create uneven running clearances in an operational engine. GM recommendations are that out-of-round cannot exceed 0.0003 inch for main journals or 0.0004 for rod journals. Similarly, acceptable taper (change in journal diameter) across the width of the journal is well under a thousandth. Fortunately, journal grinding and/or polishing can help bring journal shape within spec, but there is only so much that can be done with severely damaged or out-of-shape journals. Your machine shop will have the final say as to whether a crank with such issues will need to be replaced.

A notable wear area of the crank is the thrust surface that exists on main journal 3 (which acts to limit crank fore-to-aft movement). You should inspect this surface for excess wear and measure its width. Here you can see we're showing 1.035 inch, which is a little more than GM normally recommends; however, if the surface finish here can be polished up well, thrust bearings with thicker surfaces can be purchased to compensate for excess width.

If you're shopping for a used crankshaft, one additional item to note is corrosion. Negligent crank storage practices can cause journal surfaces to rust quickly. Though very minor surface imperfections can be polished out, this one may be beyond hope (if grinding can't do the trick either).

the surfaces, but you'll also want to physically measure some critical dimensions of the crank to see what you're up against as far as machine work.

Perhaps you're rebuilding for high-performance and have decided your needs (or desires) go beyond what a factory GM crankshaft can handle. Whether you're looking for reliability at higher amounts of horsepower, or a longer stroke for increased cubic inches, the aftermarket has you covered. A variety of quality cast and forged crankshafts are available for the LS engine family, featuring strokes from stock to well over 4 inches. Again, brand reputation is everything here, so choose wisely! If an aftermarket crankshaft is the route you're going to take, you should have a look at Chapter 5 before ordering one (in our discussion of balancing the rotating assembly, we mention certain crank-related considerations you should keep in mind).

There's a range of quality and price in aftermarket crankshafts, and some cranks can handle hundreds upon hundreds of horsepower. Like many things in life, you get what you pay for! This DragonSlayer crank manufactured by Callies is a top-quality 4340 steel forging and features a 4-inch stroke. Note that this one has a 58X reluctor ring, but 24X versions are available as well.

There are many high-strength features used in higher-quality aftermarket crankshafts, such as the large-radius fillets seen on this rod journal. Though this attribute enhances strength, keep in mind that special bearings will be required to conform to the shape of the journal edge.

A stock piston may very well be in good shape, exhibiting little wear or even carbon buildup. Conversely, check out the wear markings on this 42,000-mile Gen III piston. Skirt scuffing like this is one of the reasons that GM switched to polymer coated skirts for all Gen IV engines (as well as on stock replacement Gen III pistons—but this coating gets scratched and wears off over time, too). But no matter its visible condition, a piston is one engine part you do not want to reuse.

Pistons

The factory pistons used in Gen III/IV engines vary in exact aluminum alloy composition, but all are an excellent casting (or, in the case of the LS9, forging). Depending on the use your particular engine saw, its pistons may be in very good condition. In theory, you could even clean them up and reinstall them into your block with new rings (after just a light deglazing of the cylinder walls). In practice, however, this wouldn't make much sense, and falls more along the category of an engine repair than an actual full-on rebuild. First of all, some amount of wear to your pistons and cylinder walls has inevitably occurred no matter how few miles your stock engine ran, so though acceptable results can sometimes be had with this technique, piston-to-wall clearance will be most optimal if you hone your block and install oversize pistons.

Another reason to install oversize pistons is that most factory LS engine blocks are machined without simulating block-distorting bolt torque, and this means that although the bores of bare factory-machined blocks may spec out just fine when measured, tightening the cylinder heads and main caps in place results in them being out of round during actual operation. Problems here are especially true with aluminum blocks. Since getting the cylinders into the correct shape inevitably requires removal of material from the cylinder walls, this means clearances normally can't be proper with a standard size piston. All of this translates to the need for an oversize piston for improved efficiency and longer engine life.

Not reusing your stock pistons should be a no-brainer, and we'll discuss the block machining operations needed to accommodate oversize pistons in Chapter 5. However, choosing a set of oversize pistons definitely requires some thought. There are a staggering variety of LS pistons on the market, and the number continues to increase. Start searching around, and you'll find yourself bombarded with a variety of piston brands in an innumerable array of shapes, sizes, and alloys. After selecting the exact oversize you're going for, the first choice to make is whether you'll want a cast or a forged piston. Basically, cast pistons are formed by pouring molten metal into a mold, whereas forged pistons are forced into shape while in a semi-molten state. The latter process yields a more durable part and one much better suited to a high-performance application. (Take note that the cast vs. forged strength argument holds with other metal parts like cranks and rods, too.)

The selection process then turns to the exact aluminum alloy you'd like for your pistons, and this can get tricky. Without getting overly technical, we'll just say that cast pistons are usually available in eutectic or hypereutectic aluminum alloys, with hypereutectic being the preferred material for most applications. The most common alloys used for forged pistons are 2618 and 4032 aluminum, and it's 2618

Stroker Basics

If you're looking to up the displacement of any internal combustion engine, there are only two ways to go about it: increase cylinder bore, or increase crankshaft stroke. In the case of "boring," you're increasing the diameter of the piston, and hence, that of the hole you need to fill with fuel and air. In the case of "stroking," the pistons travel further downward into the cylinder, meaning a deeper hole is to be filled up. Either way, a bump in cubic inches is one of the best ways to realize substantial increases in horsepower and torque.

Stroking is very popular when it comes to Gen IIIs and IVs, and this goes doubly for aluminum-block engines since the ability to increase cylinder bore is so limited. Why? A minimum thickness must be kept in such a block's stock cast iron cylinder liners, and there isn't a whole lot to work with: maximum overbore for most aluminum blocks is normally in the range of 10 thousandths of an inch. Though it's possible to install cylinder liners that can accept larger pistons (think of this as the high-performance analogue to the use of repair sleeves–see Chapter 5), this process gets expensive in terms of machining costs as well as the price of the oversize sleeves themselves. Even for iron-block engines, stroking makes a lot of sense–and is especially helpful if you're looking for additional low-end grunt out of a truck engine!

But stroking an engine is not simply a matter of swapping the crankshaft out and calling it a day. Depending on crankshaft stroke and the particular block you're using, engine block clearancing may be required. A different crank will also require changes to other parts of the rotating assembly, most notably the pistons. As the piston still must reach the top of the bore during each crankshaft revolution, a stock-spec piston pin location would dictate an extremely short rod if used with a stroker crank.

Aside from physical fitment issues, stroking also puts more stress on many internal parts of an engine, most notably the engine block, so there's a practical limit here that's a function of both stroke and horsepower. Block upgrades like billet main caps or even a swap to a high-strength aftermarket block may be required. But the good news is that unlike Gen I small-blocks of years past (most of which struggled to even fit 3.75-inch strokes without starting to run into cam interference), crankshafts with 4-inch strokes will fit into most Gen III and IV blocks with minimal effort, and it's often possible to shoehorn even larger cranks in with a little work. We'll show you how to get a stroker crank to fit into your block in Chapter 7.

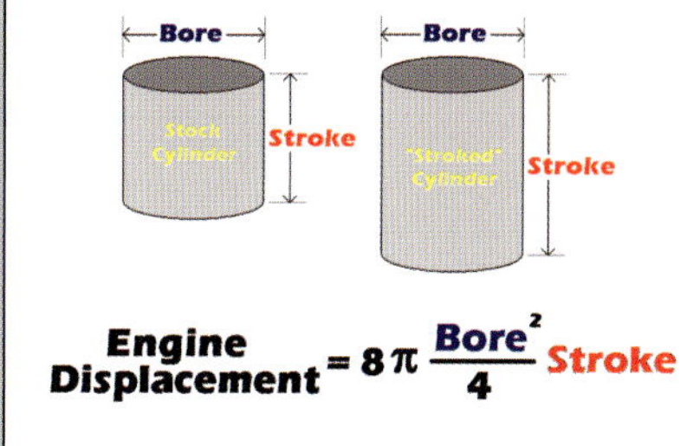

Calculating engine displacement is a simple matter of bore and stroke. You probably recognize the equation from middle school math as finding the volume of a cylindrical shape (in this case, eight of them). A stock engine has a given bore and stroke (left), while a stroked engine has a nearly identical bore (cylinder diameter) and a longer stroke (height, or more appropriately, depth of the cylinder). See the Appendix for more calculations like this.

that is considered the stronger (albeit heavier) of the two. Not all castings or forgings are the same, though, even if they are of the same alloy; so for most applications, selecting a manufacturer with a good reputation is more important than anything else!

Most high-performance applications should choose a forged aluminum piston. While much stronger than a cast-type piston and better able to withstand the rigors of higher cylinder pressures and heat, forged pistons are more expensive. The extra money is well worth it for a high-powered engine, especially when used with forced induction or nitrous oxide. This is a Speed-Pro POWERFORGED piston. As with many aftermarket pistons, this one has valve reliefs, which in this case are designed with high-lift performance cams in mind in order to increase piston-to-valve clearance (some factory pistons have them too).

Stock and mild high-performance rebuilds should choose a piston made from a cast eutectic or hypereutectic aluminum material. The advantages of this type of piston are a lower cost and reduced thermal expansion versus a forged piston. The latter translates to reduced engine noise, reduced oil consumption, and longer component life. This is a 0.010-inch oversize LS1 piston available from GM Performance Parts. Note the flat-top design, typical of many factory Gen III/IV pistons.

Keep in mind also that—along with cylinder head combustion chamber size, type of head gasket used, and other factors—the shape of the top of the piston will dictate your engine's compression ratio. Flat-top pistons (with or without valve reliefs) are the typical choice for standard to high-compression engines, but high-performance engines with aggressive camshafts will definitely want pistons with sufficiently deep valve reliefs to achieve the necessary piston-to-valve clearance. So-called domed pistons should be avoided in most cases, as while they most certainly raise compression ratio substantially, many engine builders feel that their shape tends to interfere with the combustion event (a smaller combustion chamber is considered a much better method of raising compression ratio). See the Appendix for the engine math needed to calculate compression ratio.

Dished pistons like this one made by Diamond increase total chamber volume, thereby lowering compression ratio for supercharged and turbocharged applications. As a side note here, always buy pistons and pins as a matched set and purchase the recommended pin locking rings for floating-style pins (if they're not included). This will prevent any fitment problems and help curb the possibility of pins coming loose during engine operation—a real possibility with incorrect length pins and/or the wrong locking rings.

A final note on pistons involves wrist pin style. The wrist pin (or simply pin) that connects the piston to the connecting rod can be of either a pressed or floating style. Though we'll talk more about this in our connecting rod section below, for now we'll note that almost all pistons you can buy include wrist pins, so you'll need to make sure the pin type is compatible with the rod you are using (pin diameters can differ, too). Locking rings for floating pins are normally included with pistons as well.

Piston Rings

Closely related to the issue of pistons is that of the piston ring set, comprising both compression rings and oil rings. Compression rings seal combustion gases from leaking past the piston into the crankcase, while oil rings prevent excessive engine oil from making its way above the piston and getting burned off. Type and material of piston rings vary nearly as much as pistons themselves, and the reasons for this go beyond the more obvious characteristics like piston diameter and size of ring grooves (and their vertical placement). Changes in intended engine application will dictate alterations to ring sets as well; for example, higher-horsepower applications will want rings made from more heat-resistant materials. Similarly, some high-performance builds may wish to go for a set of lower-tension rings to reduce drag and free up horsepower. Fortunately, piston manufacturers will almost always supply recommendations on rings, so this can help take a lot of guesswork out of this area.

Aside from being a wear item, piston rings must closely match your chosen piston, and you'll definitely need a complete set for any rebuild. Many different ring materials and styles are available to fit a given application and budget. Here is a full set from Sealed Power comprising (from left to right): 1.5mm moly-faced top rings; 1.5mm iron second rings; and 3mm oil ring rails and expanders.

One of the main things to decide when shopping for rings is whether you'd like a pre-fit ("drop-in") or file-fit ring set. Although oil rings are pre-sized to a given bore no matter which ring set style you choose, file-fit compression rings will be supplied with a narrow gap that must be filed wider prior to engine assembly. We'll show you how to file fit rings in Chapter 7, Pre-Assembly. If you'd like to avoid this step, drop-in sets already have their compression rings sized to a given bore diameter, but because their endgaps are designed to work in a greater variety of engine environments, they will not be as precisely tailored to your application as would be possible with a file-fit set.

Connecting Rods

Some of the most commonly (and easily) reused internal LS engine parts are the connecting rods. The vast majority of Gen III/IV rods measured 6.098 inches in center-to-center length, but there were a few exceptions to this. For example, rods used in 4.8L engines were about 0.180-inch longer, while the titanium rods used in LS7 engines measured just 6.067 inches. Keep this in mind if you are thinking about swapping in rods from a different engine.

Normally, used connecting rods will be in good condition and require very little machine work to get them ready to reinstall into an engine. If you're going to reuse your rods, about all you can do for now is have a preliminary look at them before taking them to your

There are a few different rod styles and lengths used in Gen III/IV engines, but all steel rods are a quality powdered metal forging. Certain high-output engines like the LS7 even have rods that are forged titanium! This particular connecting rod is a steel one from a Gen III car engine. Note the I-beam design.

machine shop. Have a look at the neighboring photo captions for some inspection advice.

Almost all Gen III engines used a pressed pin design that held the rod to the piston. As the pin is held into the end of the rod via an interference fit, you won't be able to remove the piston from the rod at home. Your machine shop will need to take care of this for you.

Gen IV engines (and the Gen III LQ9) used a so-called full floating pin that rotated freely in both the piston and rod (and was held in place via locking rings set in the piston at either end of the pin bore). This change was made in part to address noise issues, and floating pins are also a heavier-duty setup shared with most aftermarket rods.

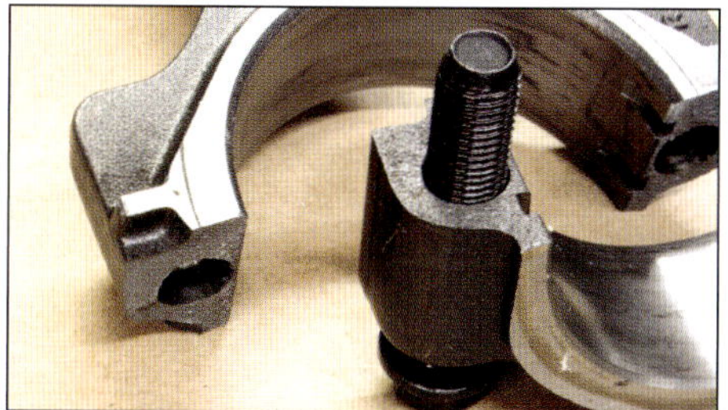

Regardless of whether you are able to separate your rod from your piston at home, you can still have a look at some critical aspects of them. One such area is the fractured (or "cracked") cap mating surface of steel LS rods, as seen here. The nature of this surface dictates that each cap is an exact match to its rod (if for some reason you lack the correct caps, your rods are garbage). Though this area is designed to look "broken," any evident imperfections–for example, from parts droppage–on either the cap or beam end of the rod mean that the rod can't be reused.

High-performance enthusiasts will be happy to know that, like factory crankshafts, GM connecting rods can handle lots of horsepower—so unless you plan on spinning a lot more RPM than stock, these rods probably do not need to be replaced on a naturally aspirated engine. This changes when you're talking about modifications like increased cubic inches. For example, while stock GM connecting rods could theoretically be used with a stroker crank, a few obstacles stand in the way of this, including compatibility issues with the large-radius journal fillets of many

Gen III/IV connecting rods use bolts that thread into the beam end of the rod to hold the caps in place. (Prior small-blocks normally used press-in studs with nuts that held the cap.) Inspect these threads for obvious imperfections. Since this is a highly stressed area, any thread damage will probably render the rod unusable (so says GM), but check with your machine shop if you're unsure.

Though bolt-related failures are uncommon for LS rods, aftermarket bolts are a common upgrade to make factory rods even tougher. More of a "factory alternative" than a true aftermarket part, these LS6 rod bolts from GMPP are the same ones used in later Gen III engines. They're available as a set and are a nice upgrade over just reusing your original bolts, particularly if you're rebuilding a pre-2000 engine.

aftermarket cranks (as well as your ability to find a piston whose compression height will match up with a stock-length rod and stroker crank). Stroker or no, aftermarket rods are also the way to go if you're looking for a piece that will be durable for higher-stress applications like nitrous oxide use.

A few pointers on aftermarket rod selection should help give you some direction. From the standpoint of engine operating characteristics, many readers probably already know that longer rods increase piston dwell time at TDC and also decrease cylinder wall piston skirt loading. On the other hand, shorter rods increase piston speed and pumping action at the expense of some frictional losses. But keep in mind that the length of rod you use is physically intertwined with your crankshaft stroke and the compression height of your piston (as shown in the Appendix). More rod length means less compression height, and for this reason, long rods can have an adverse effect on piston and ring durability in large-stroke engines. More compression height leaves more room for the rings, allowing them to sit lower from the face of the piston. This shields them from the heat of combustion and simultaneously

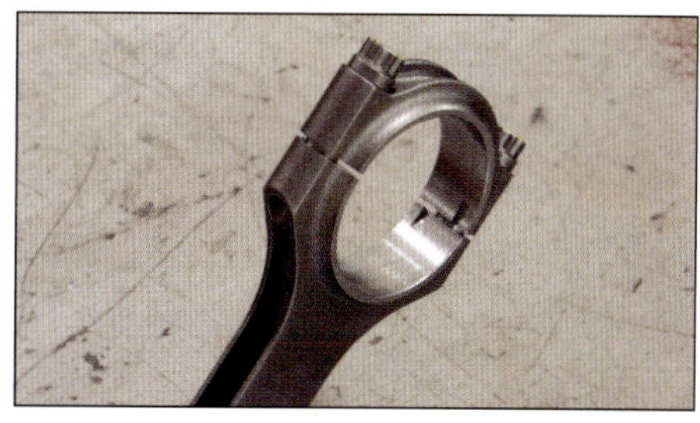

One thing you may want to keep in mind when selecting a connecting rod is that aftermarket rods with smaller-than-stock journal bore diameters are available. The advantage? It allows the overall width of the rod to be reduced and, therefore, affords more rod-to-block clearance on strokers. They of course are designed to be used with a corresponding small-journal crank. Note also that this is an H-beam rod and that it uses caps located by dowels, as opposed to the self-locating fractured design used on factory steel rods and by other aftermarket companies.

strengthens the top ring land, particularly important in nitrous and forced induction engines. For all of these reasons, it should be clear that a compromise is in order when it comes to choosing the length of rod that's right for your engine.

Lighter rotating assembly components mean less energy used to spin the engine at a given RPM, and hence increased horsepower. Though the most popular choice for aftermarket rods (lightweight or no) is steel, for the ultimate in weight savings, you may wish to choose titanium or aluminum connecting rods. These aluminum rods are destined for use in a C5R-based LS. Be aware, however, that the larger exterior dimensions of aluminum rods will require special attention to rotating assembly fitment.

Choosing the incorrect size and shape of main and rod bearings can quickly destroy your engine. Fortunately, a selection of main and rod bearings are available to fit any LS block using any combination of crankshaft and connecting rods, factory or aftermarket. Whatever bearings you choose, make sure they're from a reputable manufacturer, such as Sealed Power.

Engine Bearings

Since the LS is a cam-in-block engine, there are three different types of bearings that you'll need to worry about. They are: main bearings, connecting rod bearings, and camshaft bearings. All of them are wear items, and they must be replaced in any rebuild project.

Often considered part of the rotating assembly, main and rod bearings must be selected to match the connecting rods and crankshaft being used. Probably the most important thing to keep in mind is the shape of the edges of these bearings. While standard-style bearings are compatible with factory crankshafts, the special filleted main and rod journals of many aftermarket cranks require matching chamfered bearings. The manufacturer of your crankshaft can advise you on which bearings to use and will often sell the correct set themselves.

There are a few other factors to consider with main and rod bearings, one of which is the material that makes up the bearing. Here again, you should go by the recommendations of your crankshaft manufacturer, but we'll just mention that a wide variety of bi- and tri-layer bearings are available, and every manufacturer does things a little differently. Bearing layers may comprise steel, copper-lead, aluminum, Babbitt, and other materials depending on the company you are looking at and the intended application of the bearing (for example, level of horsepower).

A final main/rod bearing consideration is what's known as oversize or undersize. While factory crankshaft main and rod journal diameters are the same across the LS family, machine work can slightly alter the sizes of these journals (and less commonly, the bores that they ride in). So depending on the machining that must be done to your crank, block, or rods, you may need so-called undersized or oversized bearings to establish the proper running clearances for engine oil. We'll see more of this when we establish bearing clearances in Chapter 7, but for now, be aware that sometimes the need for a different bearing might not be discovered until you actually pre-assemble your engine.

Cam bearings are a little more straightforward to deal with than main and rod bearings as there are no varying shapes to be concerned with and the required inside diameter of the cam bearing is universal for all Gen III/IV engines using stock-diameter cam journals (which are the same regardless of whether you're using a factory or aftermarket LS camshaft). Also, because the physical stresses on cam bearings are peanuts compared to those experienced by the main and rod bearings, cam bearing material selection is barely worth worrying about.

Sealed Power is also a good source for cam bearings. Though selecting a set is relatively simple, the one hiccup is that LS engines used different size cam bearing bores in the block. So before buying a set, you'll need measure your block's bearing bores with a dial bore tool (or better yet, have your machine shop measure them for you). Also see Chapter 5.

If you want to get fancy with cam bearings, you might consider coated bearings. These bearings are black because of a special fluoropolymer coating that is oil-retaining and also acts as a secondary lubricant. This helps protect against dry starts and other oil-starved operating conditions. The disadvantage? Coated bearings can be expensive.

The Physics of High-Performance

Aside from increases in displacement, major enhancements in horsepower and torque output are always made via improvements to the camshaft and cylinder heads. External power-enhancing items like cold air intakes and headers are nothing without an engine that thirsts for air, and a good combination of high-flow cylinder heads and an aggressive camshaft to actuate their valves will create an engine with a truly sick appetite for atmosphere.

Folks with advanced academic degrees and decades-long careers focusing on internal combustion engines still learn about how engines make efficient power on a daily basis. But from a high-performance perspective, here's what we need to know: Cylinder head design focuses on increasing flow into and out of the cylinder, as well as tailoring how the air/fuel mix burns while it's in there. The precise shape of the intake and exhaust ports, the angle and sizing of the valves, and the contours of the combustion chamber all have a huge effect on these parameters. And here's where a high-performance cam comes in: by opening the intake valve further and longer, even more air and fuel can enter the cylinder on any given intake stroke. Increased exhaust valve lift and duration similarly allow for more complete cylinder evacuation. The end result of all this? More air and fuel to compress and ignite, and a more efficient burn once lit. This results in increased cylinder pressures, and more force pushing down on the piston–force that eventually sends torque to your tires.

We're leaving out specifics like port velocity, atomization, and chamber-induced swirl and tumble–but it's all a means to the same end. Yes, extra power is great, but there's a price to pay: the tradeoff for both larger heads and cams is almost always reduced engine efficiency at lower engine speeds and, hence, lost fuel economy and off-idle torque. What is considered an acceptable compromise is completely up to the vehicle owner.

Complex physics aside, it's all about the combination of heads and cam that will make the horsepower you want, and this is the area where things get subjective. Every expert has his or her own idea about what LS engine parts make power, but you should take any advice with a grain of salt. The most important canons to keep in mind are that not only must the cylinder heads and cam complement one another, but they must be a good choice for your engine displacement and the RPM range where you're looking to make power. To take some of the guesswork out of the selection process, some companies even offer pre-selected cylinder head and cam packages that you might want to incorporate into your rebuild.

The choice to go this route or opt to select an individual cam and head combo on your own is yours, but above all else, just make sure that companies you choose to involve are reputable ones. There is more to think about than just advertised flow numbers and lift/duration specs–high-performance heads and cams are precision-engineered pieces, and you certainly don't want a substandard part to ruin your hopes of high LS horsepower!

Cylinder Heads

The rotating assembly is what makes your engine pump air and fuel, but along with the camshaft, the cylinder heads determine how much of the mix actually enters the engine. High-flow cylinder heads have always been the key to small-block power, and the LS continues that tradition.

Many stock and moderate high-performance rebuilds may consider reuse of the factory cylinder heads, as even stock LS heads flow well thanks to their efficient port design. Initially, however, your heads probably will not look the part of a reusable item. They'll have plenty of carbon and oil residue in the exhaust and intake ports, not to mention all sorts of buildup in the combustion chambers. First impressions can deceive, however, as most Gen III/IV cylinder heads can be reused without much trouble. One of the great things is that nearly all LS heads are cast from aluminum, which is good not just because it makes them lightweight, but because it makes the heads easier to machine. This holds true not only when doing routine rebuild work like replacing valve seats, but when porting them for increased flow and horsepower potential!

Cylinder head porting is a science in and of itself, and there are many high-performance shops that specialize in this service for LS heads. If you're interested in going this route, you'll want to do some research into which of these shops are reputable (an incorrect

We went through cylinder head disassembly in Chapter 3, and now it's time to have a look at the casting. Intake ports, exhaust ports, and combustion chambers will need a little TLC to get them clean enough so you can inspect for any cracks. Don't worry, they will still look dirty when you're through, and only a machine shop's specialized cleaning equipment will be able to get them anywhere near good-looking!

port job translates to reduced port efficiency or even damaged heads). Some shops even offer CNC porting for the ultimate in consistency port-to-port. You should also keep in mind that because there are a number of different head castings used in the Gen III and IV, they vary on suitability as a starting point for this process, so don't be surprised if a shop recommends swapping your castings for other factory ones.

Though some factory heads offer the potential for excellent flow, moderate to extreme high-performance applications should also consider going with an aftermarket LS cylinder head (i.e., one not based on a GM casting). Besides offering additional flow capability, aftermarket heads typically include strengthening

As far as head checks, have a look at the castings for any visible cracks (for example, between the valve seats). Items like warpage of the casting, as well as condition of the valve seats and guides, will need the expert eye of a machinist for full evaluation. If your heads are deemed unusable, don't worry: used ones are readily available.

These LS1 valves are in decent condition; it's normal for the top of an exhaust valve (left) to be covered in a light-colored soot, as this indicates a clean-burning engine. The area atop an intake valve (right) will have an oily appearance, and engines that have seen a lot of service may also have black carbon buildup here. Though you should have a quick look at your valves for any unusual wear on the stems or seating surfaces, your machine shop will have the final say as to whether they are reusable (see Chapter 5 for more information). Stock replacement as well as high-performance valves (such as larger valves or those made from lightweight titanium) are available from both GM and the aftermarket.

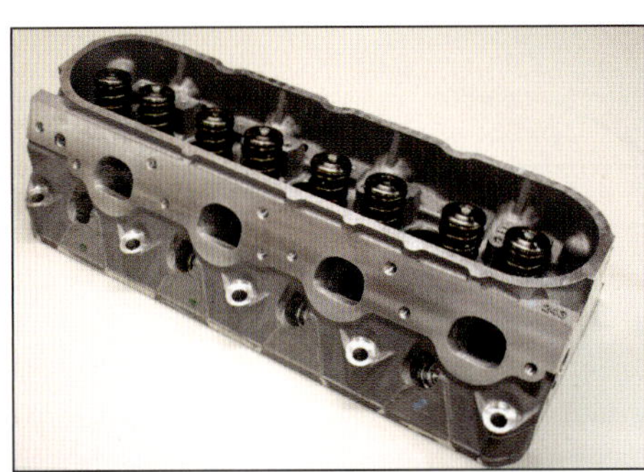

A wide variety of LS cylinder head porting shops are out there. Some will port your own castings, while others will send you already-ported units and have you return your used heads for a core charge refund, saving you turnaround time. Here's an assembled head from Patriot Performance, a popular source. You can tell this one is a GM casting from the "243" in the lower right hand corner, meaning it's an LS6/LS2 casting.

Virtually any LS head will bolt to any Gen III or IV block, so cross-compatibility of castings is excellent. There are, however, a few exceptions. Most notably, heads with larger valves and/or combustion chambers will not work with small-bore blocks. For example, this GM L92 head requires a minimum 4-inch bore to prevent bringing the intake valve too close to the cylinder wall. Aside from physical incompatibility, differences in combustion chamber size can make use of large-cc heads on smaller bore engines infeasible, as compression ratio will be reduced dramatically. This story may change, of course, if you're looking to severely drop compression for a high-boost forced induction engine!

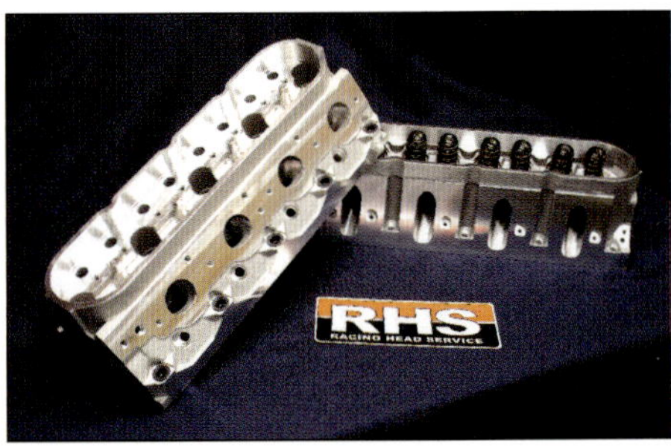

Some of the best-quality aftermarket cylinder heads available are those made by Racing Head Service (a division of COMP). Available both unassembled and ready-to-go, these particular heads are RHS's ProElite units, here equipped with high-flow 225cc intake ports. These Gen III-style heads feature a revised 11-degree valve angle, CNC machining, and other trick features for increased flow over similar factory-based castings (newer-style, rectangular-port heads are also now available from RHS). Their 0.800-inch-thick deck surfaces also make them ideal candidates for engines that will create the huge cylinder pressures typical of nitrous or forced induction.

Another great feature of the RHS ProElite head is its raised valve cover rail, which affords more clearance for aftermarket rocker arm systems when using stock valve covers. You can also see that these particular heads are supplied assembled with COMP Beehive valve springs good to 0.600-inch lift, but springs for even higher lifts and solid roller cam setups can be had as well. Also note the steel head bolt hole inserts (bottom right).

LS Cylinder Head Evolution

We talked a bit about the revolutionary Gen III cylinder head design back in Chapter 1, and although its cathedral-shape port configuration carried over to many Gen IV engines, GM began swapping over to heads with a more-rectangular port shape on higher-output engines over the first few years of Gen IV production. For this reason, it's common to refer to the earlier ports as the "Gen III-style" port and the more rectangular ones as "Gen IV-style."

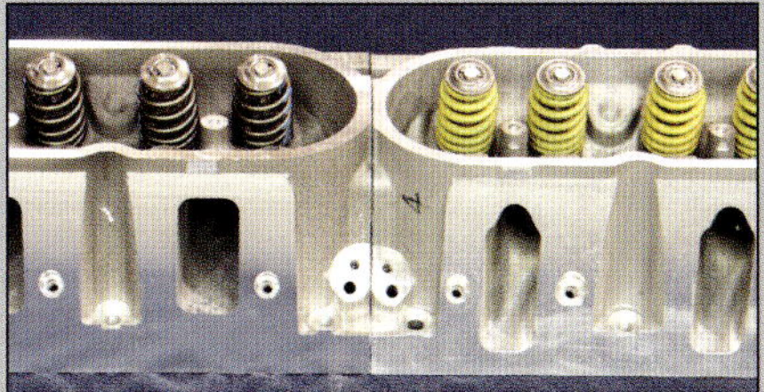

The original "cathedral" port configuration (right) stands in contrast to the more-rectangular style port that began showing up during Gen IV production (left). Notice that not only is the floor of the later port raised further from the deck surface of the head, but the port roof is somewhat higher as well. It's important to note that the two port shapes require entirely different intake manifolds.

Why the change? Ever since the cathedral-port heads were first seen on the LS1 back in 1997, many have believed this seemingly unusual port shape to be the secret to the efficiency of the Gen III head design. This is true to a degree, as although the tall port height contributes to favorable fuel injector aiming, the vertically stretched proportions also had to do with routing the port through the narrow window afforded between the intake and exhaust pushrods. The tall port height that resulted was, to a good extent, simply a necessity in order to achieve the cross-sectional area engineers were looking for.

In the ensuing years, GM threw the new Gen III architecture into the racing arena and enjoyed much success in the C5R and other racing programs. (In fact, race engineers were so successful with their cylinder head designs that they were destined to be restricted by sanctioning bodies!) GM realized that the lessons learned here could improve the designs of car and truck heads, and so tasked teams of engineers with incorporating some of this racing technology into cylinder heads destined for production engines. Without going into detailed engineering discussion, we'll just say that to achieve a straighter airflow path, both the roof and floor of the intake port would need to be brought even higher from the head deck surface than in the Gen III head. To accommodate this and simultaneously achieve cross-sectional area design targets, the intake pushrod would have to be shifted slightly to allow room for increased port width.

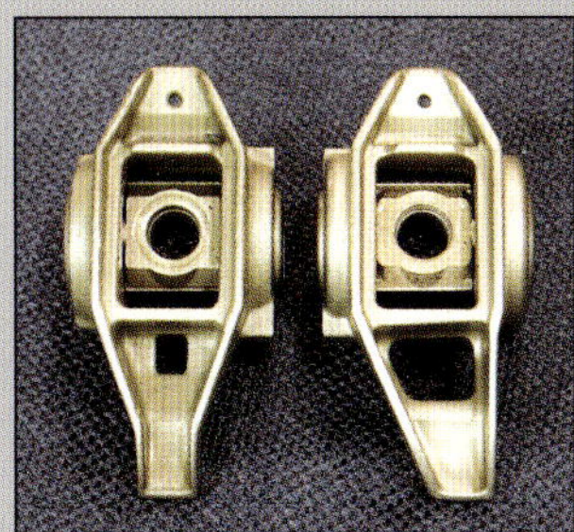

The Gen IV-style heads' relocated intake pushrod required an offset-tip rocker arm on the intake valve (right). But beware: like standard-type rockers, the offset ones were made in different ratios depending on the engine they were used in. The shape of the exhaust rockers (left) was unchanged.

The first divergent head design debuted in the Gen IV LS7, which in addition to incorporating a rectangular-style intake port, also featured a revised valve angle (12 degrees versus the original 15). This CNC-machined head was never designed with high-volume mass production in mind (and also would not work with anything less than a 4.125-inch bore), so before long, GM introduced a new head first used on the RPO L92 truck engine. This head, which also featured a rectangular port and shifted intake pushrod, kept the original 15-degree valve angle to help retain compatibility with existing production lines. The performance of this head was such an improvement over that of the original cathedral-port head design that it quickly spread to other engines in the Gen IV line.

Gen IV-style heads also came as a bit of a shock to aftermarket companies, many of whom quickly found out that the new GM heads flowed as good or better on the intake side than their own CNC-machined castings, and because the majority of GM Gen IV-style heads were an unported, as-cast design, did so for a fraction of the price! It's for these reasons that Gen IV-style heads are a popular upgrade for many Gen III engines, and a rebuild is the opportune time to swap over to a set (with the accompanying parts like a later-style intake manifold of course). Just keep in mind that because of valve and combustion chamber size factors, certain restrictions exist on which Gen IV-style heads will fit the smaller bore sizes of some Gen III engines.

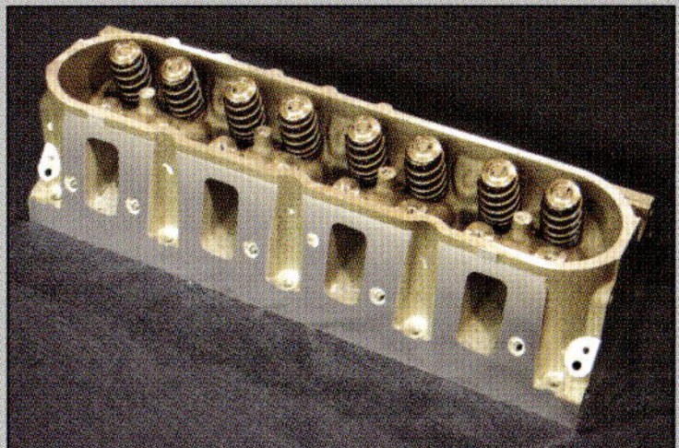

Rebuilding a Gen III engine for high-performance, but don't feel like searching for a good set of used Gen IV-style castings? You're in luck: GM Performance Parts offers the L76/L92/LS3 and other rectangular-port heads at prices so low, they will shock you. Take a look at the company website for the latest offerings.

features like thick deck surfaces and steel head bolt hole inserts. The aftermarket is continually bringing new LS-style heads to the market, and there are a lot of quality castings to choose from. Each company offers myriad options like intake port size and configuration. There's also a huge range of combustion chamber sizes available to fit a wide variety of bore sizes and achieve the compression ratios that different LS customers are looking for. With all of these options to choose from, you should have no problem finding an aftermarket head tailored to your desired application.

Valvetrain

For our purposes, the term valvetrain refers to every part of the engine involved in actuating the valves, including the timing chain assembly, camshaft, lifters, pushrods, and rocker arms, along with valvesprings and related hardware. Depending on the goals of your rebuild project, it may be possible to reuse some of these components from your Gen III or IV engine.

The lobes on this stock Gen III cam are in good condition (the lifter rollers have just "polished" them a bit). Reuse of a cam also requires verifying that the five cam journals (which ride on the cam bearings) are within spec. Using a micrometer, ensure the journals measure between 2.164 and 2.166 inches in diameter. Out-of-round should ideally be in the range of a few ten thousandths (particularly for high-stress engines like the LS7), with a full thousandth being the max for any engine.

One of the items that's easy to reuse is the LS engine's hydraulic roller-type camshaft, which probably will have experienced little, if any, wear. Because LS cams are made from steel billet with induction heat treated lobes, any damage generally will only have occurred due to a problem like a failed lifter. If you're dissatisfied with the surface of your cam's lobes, consult your machine shop on polishing options. For severely damaged lobes, cams can actually be ground (and the profile even changed if desired), but you'll need to find a very competent shop to do this and it's probably more cost-effective just to buy another cam.

To a certain extent, the same reusability goes for the rollerized-trunnion rocker arms that all factory Gen III/IV engines were equipped with. However, other items should never be reused because of their tendency to wear out over time, most significantly the valvesprings and timing chain assembly. You may be thinking, "there are no moving parts sliding against a valvespring; how the heck can it wear out?" The answer is metal fatigue, and valvesprings that had a certain amount of seat pressure

This photo notes the important difference between cam sensor placement on Gen III and IV engines. If using a Gen III block's rear-mounted cam sensor, you'll need a cam with a sensing ring located just forward of the rearmost journal (finger pointing). Similarly, if you are using the Gen IV-style cam sensor that mounts in the front cover, your timing chain set's cam sprocket will need one of two types of land-and-groove patterns: the bottom left sprocket is for earlier engine computers needing a 24X crankshaft reluctor ring (and 1X cam sensor signal), while the bottom right shows a sprocket for later engine computers that utilize the 58X crank reluctor and require a 4X cam sensor signal. Keep in mind too that you'll need a specialty sprocket (that can accommodate the required actuator) if you're rebuilding an engine with VVT.

Whereas earlier LS engines used three small bolts to hold the cam sprocket in place, later on, GM started using a "single-bolt" camshaft and sprocket design. As you can see from this photo, because the newer cams require use of a large centrally-located bolt, you will need to make sure your cam and timing set are compatible with one another (this goes for both factory and aftermarket equipment). These types of cams and sprockets started showing up around the same time that VVT debuted on other LS engines, where a so-called actuator solenoid valve replaces the central bolt.

Though early Gen III timing sets lacked some strength, later factory LS timing sets are quite robust, backward-compatible, and inexpensive. You can grab a stock-replacement chain and sprockets like this one from GMPP. Theoretically, it's possible to replace just the chain (hence their being sold separately), but this is a bad idea because the teeth on the sprockets wear out, too.

when installed will have nowhere near that after tens or hundreds of thousands of miles. Also, though not nearly as much of an issue as it is with some overhead-cam engines, the LS's timing chain stretches over time, altering and providing less precise cam timing. Though GM eventually introduced timing chain dampeners and even tensioners on many Gen IV engines, this does not alleviate the need to replace the chain.

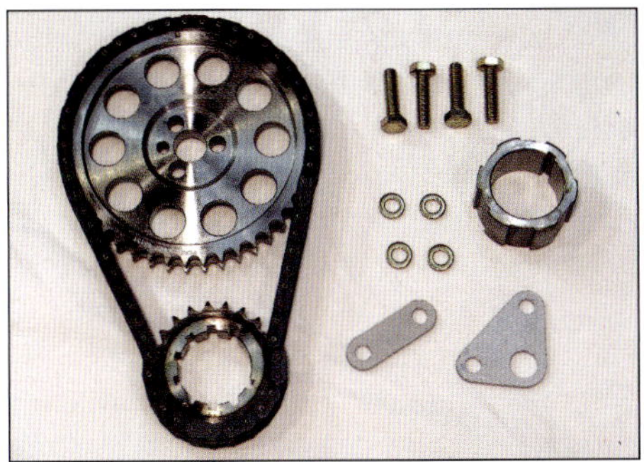

So-called double-roller timing sets are available for increased strength and reduced chain stretch over time. This one is sold by SLP and includes a spacer kit to move the oil pump forward and provide clearance for the increased chain width, and it's also adjustable for several different settings of cam advance and retard. (Other chain kits are available from SLP that include the appropriate lands and grooves for Gen IV cam sensor compatibility.)

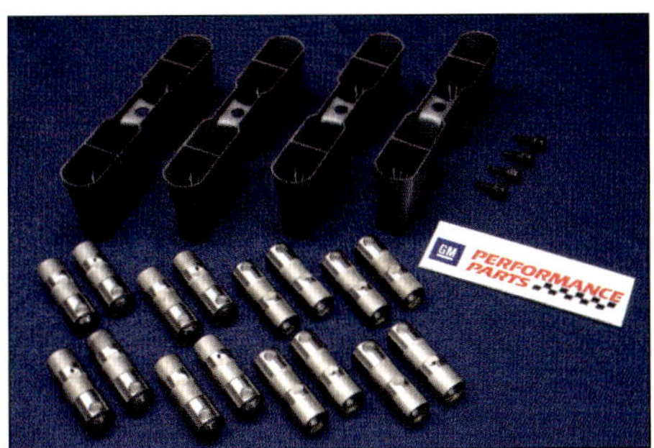

Of course, just because an item can be reused doesn't mean it should *be reused. The needle bearings inside the LS's hydraulic roller lifters will eventually wear out, so it's not a bad idea to pick up a set of new lifters like these from GMPP. You should also replace the lifter guide trays, as these items lose their tension on the lifters over time. They are what holds the lifters up during a heads-on cam swap. (Do you feel like dropping the oil pan to grab a fallen lifter? This is just one example of where you need to think ahead to possible service and/or performance upgrade scenarios down the road!) Be sure to get the correct trays for your engine, as their shape can vary slightly.*

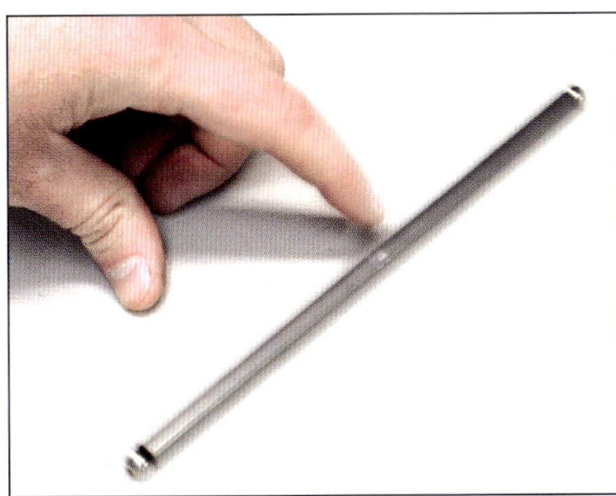

While it's also possible to reuse stock pushrods, it isn't recommended. This is because it's not unusual for factory pushrods to be bent, particularly on engines that have been over-revved. It's easy to tell if this is the case: just find a flat surface to roll them on and give them a spin. If you see even one of them wobble, you definitely should purchase a whole a new set. Federal-Mogul is a good source for stock-replacement pushrods. (See Chapter 7, Pre-Assembly, for determining the correct length pushrod for your engine.)

High-performance rebuilds often require upgrades from stock or stock-style valvetrain components. The main reason for this is that such applications will want to upgrade to a camshaft with increased lift and duration. This in turn requires aftermarket or factory performance valvetrain hardware to hold up to the increased stresses that this will induce, whether due to the more violent valve opening events, the increased engine RPM potential, or both! Stronger valvesprings are mandatory for such an application, and remember that valvesprings must be matched to your cam choice, so ask the cam manufacturer for recommendations on springs that will be a good complement before buying any. You'll also need aftermarket high-strength pushrods with a stouter cam and springs. As far as the timing set goes, most aftermarket units include additional keyways and sprocket markings to allow alterations in cam timing. Though a compromise compared to GM's VVT system, this allows some adjustment of the engine's power curve for those who wish to fine-tune (or even second-guess) a cam manufacturer's exact timing of valve events, a process that we'll see in Chapter 7.

Tougher rocker arms are a good idea to use when employing substantial increases in spring pressure and camshaft lift. Nearly all aftermarket rocker arms are also adjustable, making setting valve lash more foolproof than with the stock nonadjustable, stand-mount rocker arms (which require a change in pushrod length to adjust for too much or too little lash—some aftermarket rockers share this design too). Aside from resisting breakage, a stronger rocker design also yields more precise valve opening events, particularly at high RPM. However, we should note that the adjustable stud-mount design of many aftermarket rockers is actually a bit of a technological step backward even compared to factory rockers, which do not require guide-plates and whose stands are sturdier than any stud. Though choosing an aftermarket stud-mount rocker is usually

No matter what kind of high-performance engine you're building, COMP has an LS cam for you. This one is an XFI Xtreme Truck hydraulic roller designed for engines intended for serious towing and hauling. COMP offers a full line of Gen III/IV cams designed for any application and RPM range you might want, along with hydraulic roller lifters (a.k.a. tappets) to match.

a more cost-effective solution than going with a shaft-mount rocker system, the inherently more precise geometry of shaft rockers makes them a better choice for all-out performance. On a related note, you may also think of changing your rocker arm ratio to add additional lift at the valve, though most cams are designed to work best with the 1.7:1 ratio typical of most LS engines.

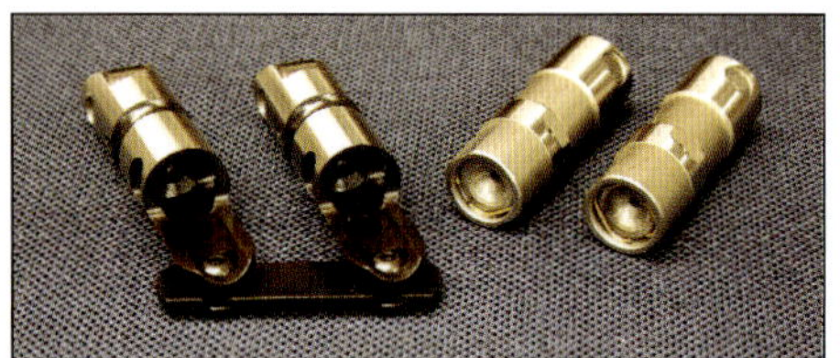

Extreme high-RPM applications may even consider installing a cam that uses solid lifters, like these LS-specific ones made by COMP (left). Though hydraulic roller lifter technology has advanced to where these types of cams can tolerate 7,000 rpm or more when paired with the correct components, the pump-up nature of hydraulic lifters (right) means it still makes sense to "go solid" if you're looking to turn, say, 8,000 rpm at the top end of a quarter-mile track! The disadvantage is the required periodic maintenance of checking and adjusting valve lash.

Stock LS valvesprings are all made from ovate wire and are an advanced tapered design that's smaller in diameter at the top. GM's LS6 valvesprings are an excellent upgrade for many rebuilds, whether it's stock or mild high-performance. You can buy a set from GMPP, and be sure to pick up a matching set of retainers and locks as well. Also pictured are replacement valve seals, which are a one-piece style that include the valvespring seat. No matter what valvespring set you go with, keep in mind that spring seats will be needed (if they are not included as part of seals). They prevent the springs from digging into the aluminum of most LS heads.

Very aggressive cams will require valvespring pressures that only a double spring design can supply. These are sold by COMP and shown accompanied by lightweight titanium retainers, spring seats and valve seals (these are a separate design), and locks along with chromemoly pushrods. Shop around, and you can find upgraded valvetrain parts packaged as a kit–nearly everything pictured can be bought together, making component matching easier.

GM rocker arms can be reused if in good condition, though it's possible for the tips (pointing) and pushrod seat to wear excessively (a bit of shininess is normal). Their internal roller bearings won't last forever, either, so unless you are really tight on cash, you should probably pick up a new set. Not all GM LS rockers are a 1.7:1 ratio, and we should also note that some aftermarket rockers look nearly identical and have higher ratios (for example, 1.85:1). If in doubt of what you have, you can measure your rocker ratio during pre-assembly or have your machine shop do it for you.

Shaft-type rockers are second-to-none when it comes to withstanding the more violent valve openings of a high-performance engine. Their stability also ensures more accurate valve events at high RPM, along with reduced frictional losses over a more conventional stud-and-guideplate rocker system. Reduced valvetrain friction equals more power and longer component life, plus all aftermarket shaft rocker systems are fully adjustable. These shaft rockers are made by COMP specifically for LS engines.

A final note on valvetrain concerns is appropriate for readers thinking of hopping up LS engines equipped with AFM and/or VVT from the factory. The AFM system can only work properly if a cam's profile has been engineered with cylinder deactivation in mind (just as an example, GM cams that were installed in AFM-equipped engines had different lobe profiles on cylinders that undergo deactivation), so if you're thinking of changing the cam in such an engine, you will only be able to use an aftermarket LS cam specifically designed for AFM compatibility. (The other option would be to disable the AFM system, of course.) VVT engines will have other cam- related concerns, so you must check with the camshaft manufacturer before using one of its products in a VVT-equipped engine. Aside from the camshaft needing the special oil passage inlet at the second journal from the front, issues arise with cam lobe profile compatibility in such an engine. Similarly, the OEM cam phaser must be upgraded when a sufficiently aggressive cam is used. In either case, there will be additional computer tuning issues to deal with once your rebuild is complete, so we advise speaking to your machine shop and tuning shop (see Chapter 9) in advance to iron out these types of issues before assembling your AFM- or VVT-equipped engine.

Gaskets

The most significant engine gaskets you'll need to get a hold of are head gaskets. Head gaskets designed for the different bore sizes of various Gen III and IV engines are available from GM and OE suppliers like Federal Mogul's Fel-Pro division. Factory-style multi-layer steel head gaskets are surprisingly durable, and as such are suitable for stock as well as mild to moderate high-performance rebuilds alike. Though often not required, more extreme high-performance rebuilds that will take on the likes of nitrous oxide or forced induction may want to step up to the extreme sealing abilities of a high-performance aftermarket head gasket. When choosing a head gasket set, keep in mind that adjustments to head gasket compressed thickness should be factored into your compression ratio equation (see the Appendix).

Although they seal decently well, the earlier-style graphite-layered steel core head gasket (left) should normally be avoided in favor of the more advanced MLS design (right). The main concern is the mess the earlier type of gasket will make should a cylinder head ever need to be removed for service or upgrade. This earlier type gasket must be used, however, with some early-style Gen III heads that had a recessed area along the outer edge of the deck surface (see Chapter 8), as the newer style gaskets will not seal properly with these heads. Both of the gaskets shown were manufactured by GM.

Other more general sealing gaskets are more straightforward to deal with. Though the LS's carrier gaskets used for the engine covers are durable enough to be reused when parts are removed for service, they should be replaced when performing a full rebuild. New front and rear crank seals to press into the front and rear engine covers are a must as well, and you should also be sure to buy new OE-style intake manifold and throttle body seals. Other miscellaneous seals you will need should have been noted during disassembly, like fuel injector O-rings and the O-ring-style oil gallery seals on the underside of Gen IV valley covers (this applies to AFM and non-AFM engines alike). All of these can be had from your GM dealer or suppliers like Fel-Pro.

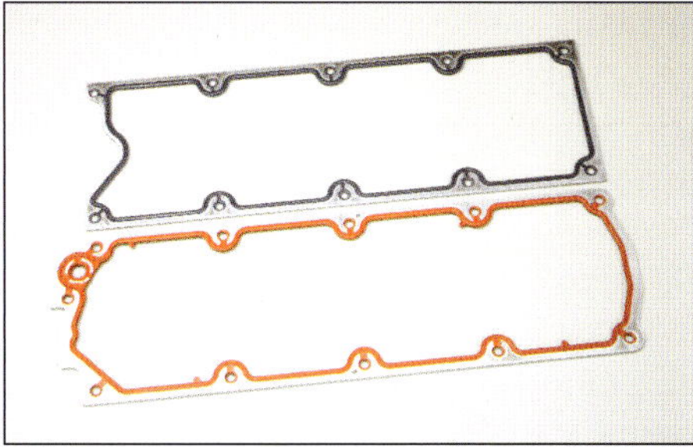

The controlled compression aluminum carrier gaskets used for the valley cover, oil pan, and front and rear engine covers are high-tech pieces indeed. They prevent gasket creep relaxation and also allow the covers to form structural parts of the engine! New ones are inexpensive, but they are not all the same (check out the difference between these Gen III and IV valley cover gaskets), so be sure to grab the right ones for your engine.

Although car engines used individual port seals between the intake manifold and cylinder head, truck engines used seals mounted in plastic carriers. The correct ones to buy will depend on whether your engine uses Gen III-style cathedral-port heads (bottom) or Gen IV-style rectangular-port heads (top).

Cometic is a good choice for high-performance head gaskets. These MLS gaskets are comprised of three layers of stainless steel for strength and corrosion resistance, with the two outer layers being Viton-coated. If you're planning on using a set of these gaskets, be aware that a special deck surface finish is required to allow them to fully seal, so let your machine shop know about it.

Fasteners

Some great news for any rebuild is that many of the factory fasteners can be reused, as there's a good chance they will be in decent condition. This is because GM used very high-quality bolts throughout the LS engine; take a look and you'll see metric grade "10.9" stamped onto the heads of many of them. Whether you're talking about engine cover bolts or main cap bolts, chances are you'll find minimal corrosion and can reuse these fasteners provided there are no issues with damaged threads.

A notable exception to the reusability rule is the LS's head bolts. While the ten smaller M8 bolts can be reused, the

Before buying a new set of head bolts, you will want to determine the style needed for your engine. Blocks cast prior to 2004 use 16 long and 4 short M11 bolts, as shown here. In 2004 and later blocks, the M11 bolt hole depths were all changed, so the M11 bolts are all of the same length (an intermediate one between the long and short seen here). The ten M8 bolts (far right) are identical for all blocks.

twenty M11 (or on some engines, M12) bolts must be discarded as they are a torque-to-yield design. This means that the first time they are tightened down, they permanently stretch slightly. Fortunately, new head bolts are inexpensive and even offered as part of a full kit (see Chapter 8 for our "GM Performance Parts" Workbench Tip). The only other LS engine bolt that is impossible to reuse is the one that threads into the crankshaft snout and secures the harmonic balancer/damper, so be sure to pick up a new one of these, too.

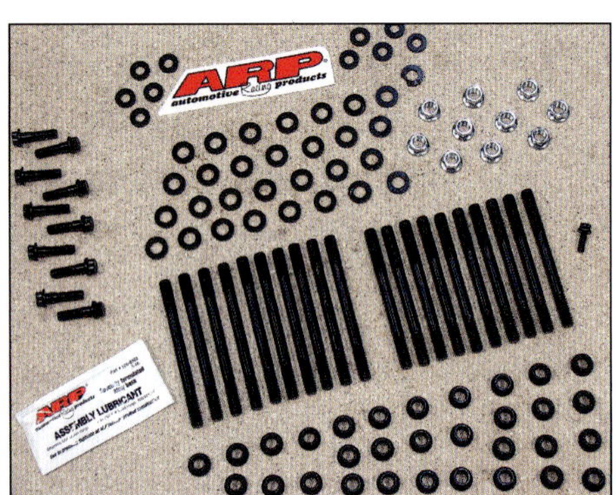

Here's an ARP main stud kit, which replaces the twenty M10 main bolts of Gen III and IV engines with studs good to 190,000-psi tensile strength. Studs also afford more precise clamping, as unlike bolts, they do not twist themselves further into the block while being tightened (the friction of which can throw off readings, particularly if you're only relying on a torque specification for tightening). They're also easier on the block threads.

High-performance applications may wish to upgrade some of the more highly-stressed factory engine bolts with aftermarket ones. ARP is by far the best source for super-strong engine bolt and stud kits, and the stock main bolts should be replaced if you plan on making a lot of horsepower. To a lesser extent, the same goes for the factory head bolts, which generally will only require upgrade if you're making lots of boost or throwing a bunch of nitrous into your LS. ARP also offers stainless fasteners for those seeking a cosmetic upgrade for bolts that will be visible under the hood.

Other Engine Parts

Putting together an LS engine requires more than just the major parts we've gone through above. Here are some pointers on some other not-insignificant parts you'll either need to get a hold of for your rebuild project or may think about upgrading for a high-performance application.

Oil pump

There were a few different styles of oil pump the factory used in Gen III/IV engines. While all were of a gerotor design, the differences between them primarily had to do with capacity, as some engines (like AFM-equipped Gen IVs) had greater oil flow needs than others. The LS7 and LS9 pumps are also different because of their dry-sump oiling system. While it's theoretically possible to rebuild an LS oil pump by replacing and/or refurbishing its internal parts, even GM does not recommend this. Besides, all engines can benefit from increased oil flow capacity, so it makes more sense to just buy a new high-flow unit. This is particularly true on early LS engines, some which were known to include less-than-adequate pumps from the factory.

The internals of an LS oil pump. Excess wear to these gears can be indicative of how your engine failed, as large amounts of metal in the oil would run through here before ending up in your filter. If you see scoring to these surfaces, you might cut open your oil filter and see what's inside! Even if these internal parts look shiny and new (as these do), your oil pump should not be reused.

Increased oil flow is good for your engine, and this is true regardless of whether you're involved in a stock or a high-performance LS buildup. This high-flow GMPP unit is an excellent choice for most rebuild projects.

Harmonic damper

It's possible to reuse the stock harmonic damper (or "balancer," if you will) from an engine that didn't see an excessive number of miles. However, these items use elastomer-type material that will eventually wear out and reduce this item's ability to cancel out harmful torsional vibration frequencies—a bad thing for your crank and bearings. (Some aftermarket dampers use other damping methods such as gels or pendulum-like internals, to varying degrees of effectiveness.) Also, any damper reuse will require it to be involved in the rotating assembly balancing process—see Chapter 5 for more information on balancing and the role of the harmonic damper.

If you'd like to reuse your harmonic damper, have a look at the section of rubber connecting the inner portion of the damper to the outer ring. Any dry-rot or other evidence of poor condition here makes reuse of the damper impossible. Also inspect the surface where the crank seal rides. It's possible to buy a sleeve to repair this area if needed, but chances are that if this surface is sufficiently worn, the unit's damping function probably isn't too well off anymore, either.

Here's yet another reason to step up to an aftermarket harmonic damper on high-performance applications: LS dampers also serve as the crank pulley, meaning underdrive of accessories is impossible without swapping the entire unit. This SLP damper is smaller in diameter and so reduces the speed of engine accessories by 25 percent, freeing up horsepower. Aside from this, it's also SFI-certified for you racers out there.

A large variety of intake manifolds were installed on Gen III/IV engines. Cross-compatibility issues arise in cable- versus electronic-throttle provisions, not to mention exterior manifold dimensions possibly interfering with installation into a given vehicle. Here we see, from left to right, an LS2 car intake (electronic), LQ9 truck intake (cable), and L76 truck intake (electronic). Some engines had a different style intake depending on the vehicle they were installed in. For example, LS2s installed in trucks had taller truck-style intakes. (Note that an LS9 or LSA has a supercharger in place of the intake manifold.)

Intake manifold and throttle body

Unless your factory composite intake has been dropped or has otherwise somehow been cracked, there's no need to get yourself a new one, that is, unless you're looking for increased horsepower! Some factory intakes are better-flowing than others, so you might consider swapping yours out for a different GM one. For example, it's popular to ditch earlier-style LS1 intakes in favor of the LS6 intake used both on LS6 and later LS1 engines. (You'll just have to keep in mind that some GM intakes were designed for Gen III-style heads, while other intakes only work with Gen IV-style heads. Even then, not all Gen IV intakes fit all Gen IV-style heads; the LS7 is a good example of an oddball manifold with limited cross-compatibility.) Even higher flow potential is available from the aftermarket, though. The best aftermarket intakes are also made from composite, but others are made from aluminum. The disadvantage to the latter is not only added weight, but the potential for heat soak affecting the density of your intake charge (and hence, robbing you of power and efficiency). On a similar note, factory throttle bodies can be swapped for aftermarket units that are larger in diameter, providing increased flow potential—though Gen IV throttle bodies were often fairly large to begin with. Changing between earlier-style, cable-actuated throttle bodies and later electronic throttle bodies is not uncommon and often fairly simple with the help of the aftermarket (and the correct ECM/PCM to enable electronic throttle actuation if needed).

Conclusion

We hope that this chapter has provided you with enough information to get you started with your LS parts selection process. Whether you're in this for high-performance or just for a stock rebuild, it's all about putting together the engine that's right for you, and doing so with the best information you have at hand. We hope that we've been able to show that because Gen III and IV engines include so much high-performance technology from the factory, you can still make a lot of power reliably on a budget; but the huge aftermarket scene also means that you can forego factory performance and build the baddest LS on the block!

Truth be told, the process of parts selection can be a very daunting one, and mentally speaking, it is easily the most trying part of any engine rebuild. It's easy to get discouraged when shopping around because there are a lot of Gen III/IV parts out there to choose from. Remember this: the final outcome hinges on your making the effort to inform yourself, set realistic goals (and a budget), and stick with your decisions throughout the stages of selecting, purchasing, and assembling your engine. Do your homework, and you'll find the parts selection process to be a very rewarding one that will pay off big time in the end.

Once you're satisfied with the parts you've selected, it's time to head to the machine shop, so read on to Chapter 5.

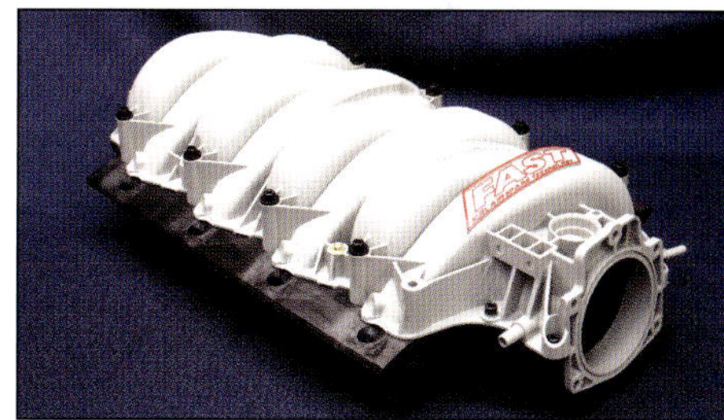

The original and still one of the best, the FAST "LSX" intake is one of the highest-flowing manifolds available on the market. It's made from polymer and includes features like molded-in bosses to make installation of direct-port nitrous nozzles easy. It's available in a few different versions with various size throttle body openings (the unit pictured is actually an earlier 90mm model that has been superseded, but it's still a great performer if you can find one used).

CHAPTER 5

THE MACHINE SHOP

Now that you've disassembled your engine and selected the engine components you'll be using (or reusing) in your new engine, the next step is to select a machine shop to work with for your rebuild project.

The type of machine shop you'll need to find is not the one turning brake drums at the back of your local auto parts store. An automotive engine machine shop is a highly specialized kind of establishment. Such a business has the trained expertise and precision equipment needed to refurbish the engine parts you may be reusing in your project, as well as perform any work that may be necessary on new parts you have purchased. A machine shop can also be a valuable resource for information, and will be happy to guide you through any step in a rebuild that you're not sure of—whether it be assisting you in parts selection, or giving you advice on running clearances. Consider the machine shop your partner for your rebuild project!

This chapter does not purport to describe each and every engine component machining process you may want or require for your project—the possibilities are virtually endless! It's merely intended as a guide to help you choose a competent shop, make intelligent choices about what machine work you need, and most importantly, ensure the work is done right.

Selecting a Machine Shop

When it comes to rebuilding your LS, there is perhaps no single task more important than finding a machine shop that knows Gen III and IV engines. A shop that does "one or two of them now and then" probably won't have the know-how to perform the machine work necessary for this particular engine family. Features like the aluminum block construction of many Gen III/IVs mean that special techniques and care are required to properly prep such an engine block. Now, don't get us wrong—you certainly do not need to find a shop that works with LS engines exclusively. To a great degree, a machine shop that works with other engine families should be considered a plus, as it shows the depth of the machine shop personnel's experience. An LS is, after all, an internal combustion engine, and so shares many basic traits with other engines, both GM and non-GM alike.

Expensive equipment is necessary to allow machinists to hold the tight tolerances needed for a durable engine, and a good shop will have a variety of different machinery in the facility. But at the same time, don't be sold on a glitzy shop with nothing but shiny new equipment. Many of the finest machines have been in service for decades, and though it takes special knowledge to properly work with Gen III/IV engine components, the tools of the trade are largely the same as they have been for decades. So don't worry if some of the machinery at the shop looks a little "experienced!"

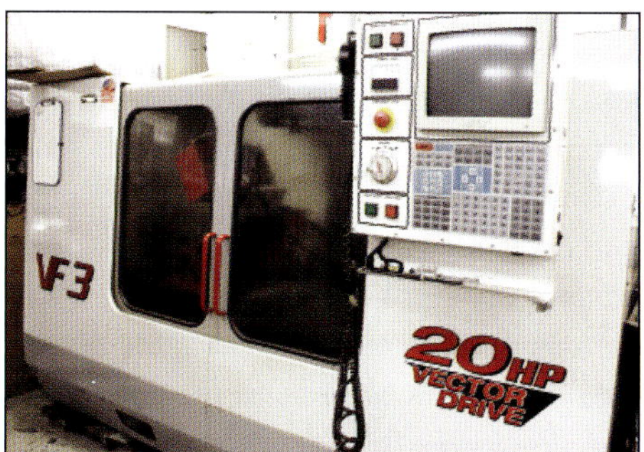

A common misconception is that if a machine shop does not use CNC (Computerized Numerical Control) equipment, it's behind the times. While these machines are very helpful in achieving the utmost repeatability in some processes that require it (for example, cylinder head porting), they are of little advantage to many others. In fact, some argue that high-quality, human-operated machinery can even have an accuracy advantage for many engine machining procedures. The main benefit of CNCing is reduced labor time for machinists, who can pretty much set up the machine and walk away. Since added productivity is offset by the enormous cost of a CNC machine, you'll probably only see this type of equipment in larger machine shops.

Machine work is not free, and while it's OK to shop around for a good price on the work you need, know that a few extra dollars spent can be well worth the investment. A capable shop may be a little more expensive than a budget one, but the added durability and performance your engine will end up with will be invaluable. On the flip side, be wary of shops that are grossly overpriced or that try and sell you much more machine work than is actually needed—use some common sense! Finding a competent, honest machine shop can be a challenge, so our best advice is to do your research and go on the advice of others.

As we go through the steps of machining, we'll highlight many important processes that need to be done in order for LS engines to perform properly. Any deviations you see from these procedures should be discussed with your machine shop—good shop personnel will happily explain to you why things are done a certain way. Though every shop has its own set of "tricks" when it comes to getting these engines to perform best, there is a big difference between a trick and an outright gimmick! Engine machinists are experts in their trade (often having attended specialized programs like those offered at the School of Automotive Machinists) and should be treated with the utmost respect, but this doesn't mean you shouldn't keep an eye out for suspicious explanations of, say, why a deck plate is not used while honing cylinders (this is a huge red flag!). On a similar note, be wary of those apt to profess their abilities while making overly disparaging comments about other shops. The best shop will be more modest and have respect for its competitors, but will of course know of some bad apples as well.

Professional engine machinists pay the same strict attention to the finest of details on every engine component that comes through the shop. Here Lou Bengivenni of LRB Performance Machine Company (Franklin, NJ) uses a file to deburr the webbing of an LS1 block, ensuring proper seating of the main caps and upper main bearing shells. Care like this is just one indication of a machinist who knows what he's doing!

Even the best intentions on a decision sometimes don't pay off. If for some reason you get through the process of dealing with a machine shop and harbor any doubt as to the competency of the work that's been done, don't worry: one of the reasons for the pre-assembly process (Chapter 7) is to ensure machine work has been performed properly. But beyond this, you have other options. Armed with the information in this chapter, you can perform more extensive at-home machine work checking using the tools and techniques illustrated in Chapters 2 and 4 (though keep in mind that the accuracy of your tools will often be insufficient to properly evaluate certain issues). If nothing else, you can always take your parts to a different machine shop for post-machining analysis.

Before Heading to the Machine Shop

Once you've selected a machine shop to work with, there are some preliminary things to take care of before packing up parts and going to the shop to drop them off. The first and most obvious order of business is to discuss your rebuild project in some detail with the machine shop owner or manager. Most significantly, this will ensure that everyone is on the same page in terms of the general scope of machine work that will need to be done. While a machinist certainly cannot know the exact condition of your parts over the phone, a detailed conversation will still help give both of you a general idea of what you're up against.

Here's another example of the type of neglect that can hold up the progress of parts machining. Even if you've purchased a pre-balanced rotating assembly and are confident you will not require any machine work on your pistons, **don't** *bring your block to the machine shop without them. Your machine shop will need to measure them, and after factoring in the proper piston-to-wall clearance, set the final cylinder bore diameter during honing.*

It's also an excellent idea to mark your parts—not just the boxes they are in—before dropping them off at the machine shop. Goodson sells helpful items like metal markers and parts tags that will help your machine shop identify and keep track of which parts are yours (decreasing the chance of accidental mix-up). **Do not** *scribe, engrave, or otherwise mark your parts in any intrusive manner—this can damage them if done incorrectly, so leave it to the experts!*

This type of talk-through will also help you understand which engine parts you'll need to bring with you to the machine shop. For example, you might be tempted to leave your piston rings at home, thinking they won't be needed until you're setting end gap during pre-assembly. But you'd be wrong: piston rings must be weighed and factored into the process of balancing the rotating assembly. Inform yourself *before* dropping off parts so that critical items don't get left behind, requiring multiple trips and delaying the progress of machine work.

Finally, before even showing up at the door of the machine shop, let them give you an honest, realistic estimate for turnaround time. Machine work involves precision processes that, like the rebuilding steps you are doing at home, cannot be rushed. Most shops will do their best to accommodate the fact that you have a vehicle you need to get back on the road, or a race you'd like to complete your engine in time for. By all means, do not simply say that you need your parts "as soon as possible!" A good shop will under-promise and over-deliver, and it all starts with your having the right attitude toward the work they are doing for you.

Block Machine Work

By far, the component of your LS that needs the most machine work is the engine block. Whether you're talking about areas that have encountered metal-to-metal wear (such as cylinder walls), or portions that have warped over time due to heat cycling (like deck surfaces), many of the block's metal surfaces will require refinishing before it can be reused. We'll go through the most important engine block machining procedures below.

Block cleaning and inspection

The most pertinent order of business is to determine whether the engine block can actually be reused at all, and if so, what machining procedures it will need. Although we went through some at-home block inspection procedures in Chapter 4, a machine shop has more

Before cleaning, the machine shop will strip the block bare of the cam bearings and all engine block plugs that you may not have removed at home. This allows the cleaning solution full access within all oil and coolant galleries. As we've seen, most plugs in the block are of a screw-in variety, but this press-in plug located at the driver side front of the block seals the main oil gallery and needs to be tapped out with a long rod from behind. (We've seen machine shops that never bother removing this plug, which is a terrible idea.)

As a used block is normally caked with sludge and grime, it will need to be cleaned before a thorough inspection and evaluation can take place. One option is this so-called jet wash cabinet, which blasts the block with a non-caustic, biodegradable alkali solution for 15 minutes at temperatures between 180 and 190 degrees F. This type of machine is commonly used to clean other engine parts as well. We should note that caustic cleaning solutions will harm aluminum and may only be used on iron blocks (iron blocks can also be soaked in a so-called hot tank if particularly dirty).

accurate equipment and the trained eye necessary to spot potential issues that need addressing. First things first, the block will undergo a thorough cleaning, and methods of doing this vary from cabinet-type pressure washers to thermal ovens that turn contamination into a fine powder (normal washing is required after this of course). Your machine shop will then inspect critical areas of the block, and at this point a good shop should give you a written estimate of the cost of machining your block, or tell you to find another one!

Once the block is nice and clean, the machine shop can fully inspect it for any cracks as well as take measurements to see what machining processes need to be performed. For example, measuring the main bearing bores determines whether a line hone is needed.

Cylinder boring

This term refers to the process of removing a large amount of material from the block's cylinder walls (generally, more than 0.010-inch). While "boring out" a block can be a great way to increase engine displacement (within the limits of what the block can tolerate), it's also a common procedure used to remove worn material from the cylinder walls and get into raw material below. It is also required whenever cylinders have a heavy wear ridge at the top of the bore. As this process only gets the cylinders roughly to size, it must always be followed by cylinder honing. Boring is generally reserved for iron blocks, as most aluminum Gen III and IV blocks do not have thick enough cylinder liners to merit the removal of so much material.

Cylinder honing

Unlike boring, honing is a very precise process that removes small amounts of material from the cylinder walls and gets them into their final shape. Generally, 0.010 inches or less are removed during honing. Honing also puts the correct surface finish into the cylinder walls, giving them the ability to hold oil lubrication and allowing the piston rings to seal properly. Again, because the iron cylinder liners used in aluminum LS engine blocks are of limited thickness, material removal from these blocks generally should not exceed 0.010 inches (though the exact amount varies by the type of block you have; your machine shop can provide guidance on this). Cylinder honing is one of the few machining processes that will be required on *all* engine blocks, and it is crucial that it be performed correctly.

When it comes to cylinder honing, a Sunnen vertical hone is the machine of choice for top machine shops. Also notice the use of a so-called deck plate.* LS blocks should always be honed with a deck plate. *By duplicating the effects of a cylinder head being bolted in place (the head gasket should be in there too!), a deck plate allows the block to undergo the same slight distortions it will experience after the engine is assembled, deformations that are easily measurable, especially with aluminum blocks. For the same reason, the main caps should also be torqued in place during this process. Far from "race-only" procedures, attention to detail like this will yield an engine with better ring seal, delivering increased durability, horsepower, and fuel economy.

Cylinder shape is only half of the equation when it comes to honing. The correct combination of stones, as well as honing machine speed and feed settings, will yield the best cross-hatch pattern in the cylinder walls for optimum ring seal and component life. Honing proceeds in stages of progressively finer stones, and for standard ring sets, a 240 grit silicone carbide stone is the preferred material for the final surfacing. For high-performance moly ring sets, a 280 or finer grit stone is normally used as a somewhat smoother finish is required.

Deck resurfacing

Over time, the deck surface of an engine block can warp, which can create problems with head gasket seal. Your machine shop will measure the flatness of the block's deck surface and determine whether removal of any deck material is necessary. Only a few thousandths of

Block deck resurfacing may or may not be needed on your engine block, depending on the surface finish of the deck and whether it's within tolerances for flatness. Any changes in deck height (beyond a simple cleanup resurfacing) may require adjustments to pushrod length as well as changes in compression-ratio-related parameters like head gasket thickness and piston design. (See the Appendix for compression ratio calculation.)

material can be safely removed from this area of the block, so severe deck warping can render a block unusable. Also, some high-performance head gaskets require a very smooth surface finish to enable a proper seal (usually 50 Roughness Average (RA) or finer); if you're using such an aftermarket head gasket, make sure your machine shop knows this and they will perform the appropriate cleanup resurfacing. Resurfacing also ensures a deck surface that's parallel to the crank centerline.

Line honing

Roundness and concentricity of the main bearing bores is very important in any rebuild. Line honing (sometimes referred to as align honing) is a common process that, although not needed on all engine blocks, is extremely important to the ability of the crankshaft to turn freely. Your machine shop will torque your block's main bearing caps in place and measure the main bearing bores for proper size and roundness. Any out-of-tolerance measurements indicate the need for a line hone, and there is little margin for error here—only about 0.0002 inches is considered acceptable for out-of-round! This spec can have a great impact on crankshaft and main bearing life.

Line honing addresses out-of-shape main bearing bores and removes material from the block and main bearing caps. Depending on the amount of material removed during this process–or the similar but somewhat more involved line boring–larger O.D. main bearings may be required, though this is exceedingly uncommon on LS engines. Line boring is generally reserved to engines where main caps are being substituted (whether with high-strength aftermarket main caps, or stock-type caps replacing ones that no longer fit properly between the oil pan rails) and is always followed by line honing.

Cam bearing replacement

The old cam bearings were removed before the initial wash, and new ones must be installed. This requires a special tool that you probably don't want to buy for your home workshop. LS engines use five cam bearings that press into place in the block. Your machine shop will measure the cam bearing bores to determine the bearing sizes you need. The reason for this is that while all Gen III/IV engines use a common size cam journal (and therefore common cam bearing I.D.), the sizes of the bores themselves vary between block castings, necessitating bearings of different O.D.s. Different sets are available to accommodate all blocks.

Shown here being removed, new cam bearings are normally installed using the very same hand tool, no press is needed. Cam bearing installation is normally the very last process performed on your engine block before it's ready to go out the door, as cam bearings can be harmed by the cleaning solutions used in the final wash.

Thread cleaning and repair

We spoke a bit about thread repair in Chapter 2, and while it's possible to perform many such fixes on your own, you may choose to leave more involved repairs to your machine shop. Whether due to corrosion, grime buildup, thread stripping, or leftover factory sealant, many block bolt hole threads can be in need of a little TLC. Without taking care of these issues, your engine assembly may be difficult or impossible. While it's true that jet washing helps clean out these holes, sometimes the process simply doesn't remove all of the grime, so it's a good idea to make sure your machine shop takes a look at all threaded holes in the block and cleans or re-taps them accordingly. Also, if you had broken off any fasteners when disassembling your engine and haven't extracted them on your own, your machine shop can remove them for you—as well as perform any other thread repair that may be necessary.

Other processes

There are a plethora of other machine work processes your shop can perform on your engine block, and entire volumes can be written on the topic alone. These may include high-performance modifications like drilling and tapping your block for the installation of a timing

Resleeving is a process used when a cylinder wall is too thin or too badly damaged to be cleaned up with an oversize bore or hone. The procedure is more common for aluminum blocks because (again) their iron cylinder liners dictate limited oversize allowance. However, even the most competent shop will tell you that there is a lot of margin for error in resleeving and that the time commitment will result in a rather expensive job. Also meriting brief mention is the high-performance modification of installing "big bore" cylinder sleeves (a process that has fallen out of favor as affordable large-bore engine blocks like the GMPP LSX have become available); not only can the process be cost-prohibitive, but some question the structural integrity of an engine block so modified.

chain dampener (if you didn't have one already), machining the block for use of special head gasket sealing O-rings, or the fitting and installation of aftermarket high-strength main caps. Or they may include repair processes like installing lifter bore sleeves for worn lifter bores. Shop personnel may even suggest procedures they feel are necessary to bring an otherwise-useless engine block back from the brink of "beyond hope." Examples of these include cylinder resleeving and engine block welding. But be wary of extensive block repairs; aside from some of these repair scenarios being questionable in mechanical integrity, their cost can often exceed that of simply buying another block.

Final cleaning

After all machine work has been performed, there will be substantial leftover metal debris throughout the block. Your machine shop will give the block a final thorough cleaning to make sure all areas—particularly the oil passages—are free of contamination. Most shops have a large tank of mineral spirits that they will submerse the block in, and will perform a thorough brushing of all passages. Your machine shop should leave out all passage plugs for you to install at home during final assembly. Note that this final cleaning process does *not* negate your need to perform the at-home cleaning procedures detailed in Chapter 6!

After the final cleaning, your machine shop should deliver your completed block to you in a sealed bag to keep dirt, moisture, and other contaminants away. They'll also have sprayed key areas of the block with some temporary rust protectant so that it can safely sit in storage until you're ready to work on it.

Crankshaft Machine Work

If you're purchasing a new crankshaft, it should require little (if any) machine work, as they are generally shipped ready to assemble. The only exception is that, as with a used crankshaft, a new crank must be balanced along with the rest of the rotating assembly. This process will be discussed in detail momentarily.

On the other hand, reuse of a factory crankshaft will require varying amounts of reconditioning. We discussed some examples of crankshaft journal damage in Chapter 4, and depending on its extent, a machine shop may be able to grind your crank journal(s) down past the damage. This will leave a clean surface that can be polished to like-new specs and used along with an undersized (smaller I.D.) main or rod bearing. However, there is only so much material that can be taken away before the structural integrity of the crank is compromised, so some cranks may not be salvageable. Some GM literature actually recommends no grinding of LS crank journals, but skilled machine shops do it successfully all the time!

Crank grinding can also serve purposes other than mere reconditioning. For example, one may wish to increase cubic inches via a so-called offset grind. By removing more material from one side of the rod journal than the other, crankshaft stroke is increased slightly. Although the cubic inches yielded are small compared to installing a stroker crank, it may be worth it on some applications—particularly if your crank journals need to be ground anyway. As with any "stroked" engine, keep in mind that an offset ground crank may require changes to other parts of the rotating assembly (see the Appendix). Conversely, offset grinding can be performed in order to *avoid* changes to other parts of the rotating assembly. For example, if you're using a block that has had a substantial amount of material removed from its deck (uncommon for LS blocks), offset grinding can be used to *decrease* crankshaft stroke slightly, thereby maintaining the proper location of the piston face at TDC. A word to the wise: if you're going to attempt any of this, make sure the machine shop (or specialized crankshaft shop that they send work to) is a competent one; done incorrectly, crank grinding of any kind can leave you with out-of-phase piston movement, a weakened crankshaft, or other problems you don't want to deal with!

Because crankshaft journal grinding reduces or eliminates the hardened outer layer of the journal surface of some aftermarket cranks, these cranks must be heat treated after grinding. There are a few different methods of doing this, and this is a crank fresh from being nitrided (you can tell from the dull finish of the journals; it's not an LS crank, but you get the idea). Some more work needs to be done before it's ready for use. Induction hardening is another popular heat treatment process, and one that generally achieves a greater depth of hardened material. We should note that factory nodular iron LS cranks are not heat treated from the factory, but can be subjected to such a process for a more durable journal surface finish.

Before polishing, it's common practice to chamfer the openings of the oil passages in the crank journals for increased oiling efficiency to the bearings. We should note that many machine shops do not perform crankshaft machine work in-house, but rather farm it out to experts. A crankshaft is one of the most precision pieces of the entire engine, and it takes specially trained hands to prep one properly!

A typical crank journal polishing process involves spinning the crank in a special lathe while running three different polishing bands of varying coarseness over the journals. The result: super-smooth rod and main journals that will last many thousands of miles.

Crankshafts with journals in good condition represent a much simpler scenario, as they'll simply need to be polished to achieve an acceptable surface finish. This isn't to say such a crank won't have other issues that must be addressed. One common crank problem is lack of straightness (or "runout"—the centerline of the crank is actually bent such that it would wobble while spinning; small amounts of this are acceptable). This may sound incurable, but believe it or not, the process of straightening a crank is often very straightforward, involving little more than clamping it into a special crankshaft straightening press and hitting key locations with a special chisel, thereby relieving the internal stresses causing it to bend.

Finally, there is one part of the LS crankshaft that was distinctly absent from past small-blocks, and that is the reluctor ring. Although sometimes this item doesn't require addressing before the crank can be reused, on occasion the reluctor ring is out of shape and won't allow the crankshaft position sensor to generate a usable signal. Your machine shop can remove and replace your reluctor ring for you (sometimes it's a non-issue as it can literally fall off of the crank after heat treatment!). No matter what, don't ignore your reluctor ring and assume that it will work, because you certainly don't want to finish assembly of your engine only to find that it won't fire up! (We'll also show you how to check reluctor ring runout in Chapter 7.)

Connecting Rod Machine Work

As with new crankshafts, most new aftermarket connecting rods are shipped ready to use and require almost no machine work (other than balancing). We say "almost" because depending on whether you are using a pressed- or floating-style piston pin, your machine shop may need to install your pistons onto your rods. Regardless of whether this is necessary, you should also have your machine shop take a look at your aftermarket connecting rods to inspect for obvious manufacturing defects.

Happily, the case is similar with used Gen III/IV connecting rods, as they normally require surprisingly little machine work. The forged construction of these rods means they are very durable, and therefore, normally in excellent condition even after many thousands of miles of use. The first order of business for your machine shop is to scribe your rods and

Goodson makes this tool to properly index and install a new reluctor ring onto the back of an LS crank. Because the process requires a press and/or heating the ring to high temperature, you should probably let your machine shop take care of this for you. If you'll be doing it at home, be sure to pick up a set of high-heat gloves and a temperature indicating crayon from Goodson as well! (In case you're wondering, the old ring comes off easily with a hammer.)

Reuse of factory Gen III rods (LQ9 engines excepted) will require a press-type piston pin. A hydraulic press is standard machine shop equipment to press used pistons off of connecting rods. Later, new pins are installed with a new piston, but a press is not normally used for this (the end of the rod is instead heated so that it expands and allows the pin to slide in—a procedure that requires a quick and skilled hand!). This is better considered rod machine work than piston machine work since, after all, the pin is pressing into the rod (and rotating freely in the piston).

caps so that each pair is uniquely identified. Though the fractured design of factory steel rods means that each cap is a unique fit, some fracture patterns may be similar to the point that it's difficult to tell which cap goes with which rod.

After a good cleaning, your machine shop will then perform a series of checks. Rods will be inspected thoroughly for any imperfections, including twisting, damage (such as nicks or cracks), and proper fit of the cap. At the top end, rods using a floating piston pin have a bushing that may need to be replaced. Earlier-style rods can also be bushed for use of a floating pin if the needs of your build dictate, though this process can be more costly than just acquiring a new set. Repair of LS connecting rods is limited, and generally rods that don't pass muster should be thrown away. Used rods should be easy to get a hold of, and you can also buy several different brand-new Gen III/IV connecting rods from GM Performance Parts.

A few final cautionary notes are in order with regard to aftermarket rods, particularly when assembling a large-stroke engine. If you did your homework during parts selection, it's likely that your rods will fit just fine with your rotating assembly and within your block. However, with certain high-end, high-performance applications, your machine shop may forewarn you of minor fitment issues you may be likely to encounter. You should not be overly concerned about this, as it is normally reserved to exotic applications like the use of aluminum connecting rods; but if there is any doubt as to whether your rods will need to go under the knife for a little extra machine work, your machine shop will tell you to pre-assemble first so that any issues can be identified and addressed *before* you spend the money to have your rotating assembly balanced.

Your connecting rods will be marked using a method that won't create a stress riser on the surface of the rod (or ruin the roundness of the bearing bore), and scribing is perhaps the best way to go about this. This marking process is to the benefit of both you and your machine shop, preventing mix-ups on their end as well as helping you out later during assembly.

Piston Machine Work

We stated in Chapter 4 that pistons should not be reused in a full engine rebuild. Aside from being mass-matched during the balancing process, your new pistons will require few or zero machining operations to get them ready to install.

As with many other internal engine parts, used connecting rods will be washed in a mineral spirits solution and thoroughly scrubbed of any contaminants. Special care is taken with the bolt hole threads as any sludge here could interfere with proper bolt tightening, resulting in possible disastrous failure. Again, any physically damaged threads may render the rod unusable.

Applications that may require some piston machining are generally reserved to higher-end custom applications. This mostly includes machining of the face of the piston (often called flycutting or milling) either for addi-

Your machine shop will tighten the rod cap in place and measure the roundness of the rod bearing bore using a precision measuring machine like the one seen here. Out-of-round beyond a few ten-thousandths indicates the need for a replacement rod—any machine work to correct the problem could be risky and expensive, and the fractured design of the cap mating surfaces typical of nearly all LS rods prohibits the practice of "cap cutting."

Although uncommon, rods may require some machine work to properly fit with other parts of the rotating assembly, as was the case with this 427-ci, C5R-based build. Here a portion of an aftermarket aluminum rod was found to interfere with a crank counterweight during pre-assembly, requiring the rounded edge shown to be machined off. A good machine shop will advise you of the potential for any such problems so that they can be dealt with before balancing.

This is one reason that you should pay attention to crank counterweight specs when shopping for a crank—aftermarket cranks are often sold "rough" balanced to a certain maximum bobweight value, and you should make sure you're not going to go over this value or you risk the added expense of heavy metal insertion. The same can go for use of some GM cranks, especially when used in applications with larger displacements (i.e., bore sizes) than the crank originally was intended for.

In addition, there are a couple of items that aren't actually inside the engine that nonetheless can affect engine balance. We're referring to the harmonic damper and the flywheel (or flexplate). Though all good aftermarket dampers and flywheels are neutrally balanced, we mentioned in Chapter 3 that some Gen III/IV harmonic dampers had small balance weights installed into their perimeters to account for production line variation. In a way, this really did make them "balancers," which normally would be a misnomer! Though the amount of mass used here pales in comparison to the "externally-balanced" 400-ci small-blocks of the 1970s (which had large counterweights on both the damper and flexplate), these weights are still significant enough to affect engine balance. Therefore, if you're planning on reusing your factory harmonic damper, you'll want to take it along with you so that the machine shop can factor it into the balance of your rotating assembly. The best thing to do is to have the machine shop add, subtract, or otherwise adjust these weights so that your damper is neutrally balanced. We also recommend checking this out for factory flywheels or flexplates if they are being reused. Neutrally balancing these components just makes things easier, particularly if a flywheel or damper ever needs to be replaced down the road, and also since most stock dampers have no alignment keyway!

Here's an example of where the factory harmonic damper balance weights mentioned in the main text might go. Just so we're all on the same page, balancing does not negate the need for a functional harmonic damper! Dampers absorb energy from crankshaft torsional vibrations (most significantly, from the firing of each cylinder), limiting distortion of the crank. These forces exist independently of the forces that would be created by an unbalanced rotating assembly—you still need a good damper, no matter how good your balance is!

Cylinder Head Machine Work

The amount of work you'll need your machine shop to perform on your cylinder heads depends completely on your intended application. If you are planning on installing a set of new head castings, this represents a much easier scenario; many aftermarket and factory high-performance LS heads come fully assembled and require no machine work at all. Heads that do not come fully assembled ("bare" heads) can be sold in a variety of stages of preparation, needing as little as the installation of valve springs or as much as the installation and machining of valve seats and all other hardware.

If you plan on reusing your head castings, however, the first order of business (other than ensuring the castings are not cracked and can be reused) depends on whether you plan on having a porting job done to increase their flow potential. We talked about this in some detail in Chapter 4, so just know that if your heads are being ported, this is usually the first bit of work that gets done. Because this process is very specialized, your machine shop may send your heads out to an expert to have the porting work performed (whether it's a traditional port and polish or a full CNC'ing).

Regardless of whether your heads are undergoing porting, they will need to have their valve seats reconditioned. Although seats can be replaced if damaged, this is normally not necessary, and factory LS seats are high quality since they are hardened (they're also fine for most high-performance applications). The typical machining procedure for all seats is a 3-angle valve job, the standard being 30, 45, and 60-degree angles.

Similarly, intake and exhaust valves themselves must be reconditioned. Standard valve reconditioning involves refacing (grinding) valves on the 45-degree cut that contacts the valve seat. The width of this 45-degree cut affects valve cooling: too wide or too narrow, and the valve will overheat. The widths and angles of the cuts on a valve are all up to a given machinist, but any "tricks" to increase flow should of course only be used if they've stood the test of time on

While the standard 3-angle valve job is all that is needed for stock and most high-performance rebuilds, deviations in the angles of the three cuts are possible to increase flow, but often at the expense of component longevity (for example, 38-45-58 or 35-55-68, the latter being a strict race-only setup). Experienced shops like LRB know that Newen cutting equipment is top-shelf stuff and won't use anything else for valve seat machining.

Here, a valve is refaced on a special grinding machine. Since this process spins the valve at high speed, it will also reveal any issues with valve stems that are not straight. This is not uncommon on LS engines: from this set of 16 used LS1 valves, 5 intake and 3 exhaust valves were bent! Fortunately, replacement valves are readily available from the likes of GMPP, Federal-Mogul, and other suppliers. It should also be noted that some valves are not capable of being refaced at all, so don't be surprised if your machine shop has to acquire new ones for you.

Once the valves and their seats have been machined, valve seat quality can be verified. Layout fluid is applied, and the valve inserted and spun by hand while pressure is kept against the seat. Once removed, the valve should look shiny on the center of its 45 degree surface, with consistent lines of fluid on either side that are somewhat narrower on the outer edge all around the valve's circumference.

Head resurfacing may be performed on the very same machine used for the block's deck surface. In a simple cleanup resurfacing, around 0.010-inch or less is normally taken off. This is another example of where CNC equipment is not a necessity; for routine head work, tried-and-true machines like this one will do just as good of a job!

other engines! Some valves are not reusable because the stems themselves are worn, often the simple result of stem-to-guide friction, but traditionally also possible because of (believe it or not) valvestem seals. This is unheard of on factory Gen III and IV heads because GM uses good quality seals on all LS engines (be sure to select Viton or similar valvestem seals for your rebuild—Teflon seals cause substantial wear on the valvestem). Beyond this, there isn't much to say about valve machining, save that worn valve tips from a faulty rocker arm, inadequate oiling, or simply an excessive number of miles of use can be ground down to a smooth finish.

Stock Gen III/IV valve guides are of a press-in variety that drive out from the bottom and are inserted from the top (shown) using a guide driver tool. If replacing guides, they are either reamed (iron) or honed (bronze) to size after installation. As a side note, worn guides should always be replaced, never knurled. Your guides will not last long if subjected to this "stopgap" process, and if your machine shop has the audacity to suggest it, you should definitely "head" elsewhere!

The other items that wear out in a cylinder head are the valve guides. Stock GM LS valve guides are long-lasting iron and often do not even need replacing on heads from lower-mileage engines. Your machine shop will verify valve stem-to-guide clearance—this is important not only for valve and guide longevity, but because excessive clearance will increase oil consumption. A typical specification for stem-to-guide clearance is 0.0017 to 0.0020-inch, with variation depending on whether you are talking about intake or exhaust valves and the type of guide material used. Some high-performance applications may wish to use bronze valve guides. A favorite of aftermarket cylinder head

When milling heads to increase compression ratio, your machine shop will use a burette to determine the combustion chamber volume. This process is sometimes known as "cc'ing the head," since combustion chamber volume is normally measured in cubic centimeters (cc).

Regardless of whether you'll be assembling your own cylinder heads, you might want your shop to at least measure your valvesprings and check that they deliver the proper spring rate and seat pressures. Specs can normally be verified with those listed on the instructions included with the springs (or shown on the outside of the box).

manufacturers, bronze guides are claimed to reduce friction and provide superior heat dissipation from the valvestem, but many machinists still feel that they will not last as long as iron. Unless your existing valve guides need replacement anyway, it's probably not worth the expense of swapping over to bronze on most applications.

Cylinder head resurfacing is a very good idea for all heads, regardless of miles of prior use or intended application. A resurfacing machine is used to correct for small warpage and surface imperfections, and a machine with quality cutters will also yield a fine finish on the head deck surface (as with the block deck surface, this is a boon for high-performance MLS gaskets that often will not seal properly with anything less). Resurfacing machines are also used to mill heads to reduce combustion chamber volume and thereby increase compression ratio on high-performance applications. There are limits to how much material removal different heads will tolerate, however, and if considerable milling is performed, the machine shop must also cut the head's intake flange to avoid misalignment with the intake manifold.

If you'll be assembling your own heads, the work of your machine shop pretty much ends here. However, if you don't mind spending a few extra dollars, you can save yourself some time by having the machine shop install valve stem seals (this is strongly recommended if using early-style "2-piece" seals that are separate from the spring seat) as well as springs and related hardware, so that basically all you will need to do back at home is bolt your heads on. We'll discuss head verification and assembly procedures in Chapter 7 and Chapter 8.

Other Machining Processes

We've just gone through the major types of machine work you're likely to need on your engine parts. However, a slew of others may be recommended by your machine shop based on factors like the condition of your used engine parts and the characteristics of your particular application. It is not possible to mention them all here, but don't be surprised if your machinist starts throwing terms like "Magnaflux," "Meta-Lax," and "parts demagnetization" at you. The internet and other CarTech books are great sources of information, so familiarize yourself on such items if you are not satisfied with any explanation your machanist provides.

Final Machine Shop Notes

At this point, you should have a great appreciation for the wide variety of engine component machining processes your machine shop can take care of for you. Many of them are involved and complex, and the types of work needed (and extent to which each of them must be performed) all depend on the state of your parts and your desired end product.

Remember, if you come across any information that conflicts with or wasn't mentioned in this chapter, don't be afraid to do your research and question your machine shop about any given machining process. An educated consumer is a good consumer. But at the same time, don't lose respect for the machinists, who have technical training and many years of specialized experience. Help them to help you by taking a positive attitude toward their work, and trust that any problems that might arise will be resolved in due course.

You may be excited once all of your freshly machined parts are in your hands, but don't get ahead of yourself—final engine assembly is still a little ways off. With all machine work having been performed, it's time to move on to Chapter 6, Component Cleaning and Preparation, followed by Chapter 7, Pre-Assembly Procedures.

Component Cleaning and Preparation

Don't just install your engine components right out of the box! Whether you're talking about new parts, or parts that have been refurbished by your machine shop, they are almost never intended to be installed right away. One dirty item can contaminate your entire engine!

OK, so you have all of your parts in hand, whether brand-new or freshly machined. We're definitely making progress toward enjoying a fresh Gen III/IV engine—but let's fast forward a moment. Say you've got your rebuilt engine up and running, and everything is going great. You're enjoying the new-found power of your LS-equipped ride, and wake up every morning anxious to drive it (or race it, as the case may be). Now imagine that you check the dipstick and discover you're a little low on engine oil. But instead of adding some fresh oil from a sealed container, you instead grab a handful of dirt and throw it down the fill cap. Such an act may sound far-fetched, but in effect, this is exactly what you'd be doing if you didn't take the time to properly clean all engine components prior to assembly!

There are a few reasons why this analogy holds. First and foremost, although most machine shops do a very meticulous job of cleaning all of your freshly machined components before you pick them up, there's always the chance that something has slipped the trained eye (and parts washing equipment) of the technician. At best, leftover metal shavings will be small and cause scoring on parts of the engine, reducing their useful life. At worst, a large enough chunk of material left inside the engine block can easily clog your main oil gallery, robbing the engine of oil pressure and destroying it within a matter of seconds. Ultimately, you are the one responsible for ensuring all engine components are 100 percent clean before assembly.

Moreover, a metal component that appears outwardly clean and shiny may hide a dirty little secret—literally. Bits of dirt and metal are often too small to be seen with the naked eye, but no matter how small these particles are, they can still cause damage. Probably the best example of this occurs with cylinder bores. During the honing process, tiny metallic particles become embedded in the cylinder wall, and they can only be fully removed via meticulous scrubbing using the correct chemicals. Your machine shop may have done some of this for you, but don't assume it was done to its full extent.

Finally, even if you're using brand-new, out-of-the-box parts that require no machine work (for example, a set of fully-assembled high-performance cylinder heads), there's no guarantee they are clean, either. Parts may have been sitting on a shelf collecting dust before they were boxed up and sent to you, and the same goes for the box they were shipped in! Think about all possible sources of contamination along the chain of custody of any item, and you'll quickly realize that when it comes to engine components, virtually nothing is truly "ready to install."

Even brand-new parts are often not clean. For example, you can see that peeling the packing tape from the deck surface of these aftermarket cylinder heads reveals substantial metal shavings in the coolant passages. **Never** *assume parts are clean, even if they appear so outwardly!*

A little bit of effort and elbow grease during the cleaning process will pay huge dividends in LS engine longevity, not to mention efficiency and horsepower. As time consuming as this process may be, it is *not* to be taken lightly!

Cleaning Supplies

The good news about engine component cleaning is that few specialized supplies are needed to perform the process properly. In fact, you probably have most of the items you'll need to clean your engine parts, and a lot of the rest are readily available in your local hardware store. Beyond these run-of-the-mill supplies, the few more specialized items are readily available from engine builder suppliers like Goodson. We'll show you everything you need here.

Solvents

You'll need substantial amounts of a good general cleaning solvent. This will be used both to soak dirty parts as well as to wipe surfaces with. Probably the best such solvent is mineral spirits, an item that is readily available in the paint section of any hardware store. This is the same stuff most machine shops have circulating through their wash tubs. It's not a bad idea to have some carburetor cleaner (the sensor-safe type) and brake parts cleaner around, either—they'll do a similar job for surfaces where a spraying action is preferred. The important thing to keep in mind when selecting general cleaning solvents is to find types that will evaporate and leave no residue.

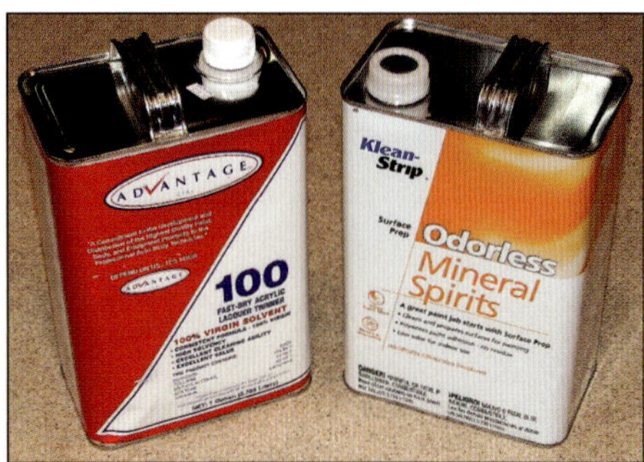

Used mainly in paint preparation, mineral spirits is inexpensive and can be found almost anywhere. You'll go through a lot of this stuff, so get at least one large container of it. A stronger solvent you may wish to try on particularly dirty parts is 100 percent virgin acrylic lacquer thinner, but this stuff is a lot nastier, highly flammable, and should be used sparingly and with extreme care.

Carburetor cleaner and brake parts cleaner will come in handy when spraying into tight passages and bolt holes. So-called electric motor and parts cleaner is another option.

Other cleaning agents

There are a few other liquids you'll want around to help you in cleaning, some of which aren't usually thought of as cleaning agents. Many of you probably know that WD-40 is very good at dissolving grease, rubber, and other substances, and it will come in handy for this very reason. Also pick up at least a quart of automatic transmission fluid. This may sound strange to you, but the detergent properties of ATF make its cleaning ability second to none when it comes to removing leftover metallic particles from your cylinder walls. Engine degreaser (the stuff used to spray down and rinse off a dirty engine compartment) can be used to clean heavily soiled intake manifolds and other non-metal items, but *must* be avoided on internal engine parts.

Believe it or not, automatic transmission fluid is an excellent cleaning agent for metal. WD-40 easily dissolves many types of gunk and is also an excellent metal protectant, so it will come in handy in a variety of instances. Not shown is engine degreaser (often known simply by the "Gunk" brand name), which should only be used on non-metal items.

Cleaning for Pre-assembly vs. Cleaning for Final Assembly

This chapter is written with a mind toward cleaning all engine parts completely and thoroughly. The reasons for doing this before final assembly should be clear from the introductory text of this chapter. But to avoid any surprises on your part, we should tell you up front that you'll need to perform most or all of these cleaning procedures twice: once before pre-assembly (Chapter 7), and again before final assembly (Chapter 8). Why the double hassle?

A quick glance at Chapter 7 should make you well aware of the kind of parts contamination that can occur during the pre-assembly process–just have a look at piston ring fitting, or better yet, block clearancing for stroker cranks. Since parts will get very dirty and oily during pre-assembly, you might be tempted to forego the first cleaning process and say, "The heck with it, I'll just clean it all once before final assembly." This would be a grave mistake. Many parts contact one another during pre-assembly procedures, so one contaminated part can quickly turn into five contaminated parts. Worse, leaving parts dirty during this process could cause actual component damage. For example, not properly cleaning a crank before setting it in the block and torquing the mains to check bearing clearances could embed dirt into the bearings, making them useless.

Think of it this way: make your parts as clean as you possibly can right now to get them ready for pre-assembly. Then, after pre-assembly, make them even cleaner than that!

Containers

You will want containers of various sizes in which to soak and wash parts. Either plastic or metal containers will do, but you'll need to get them nice and clean before use—mom's old baking trays will work, but they cannot be rusty. Feel free to improvise with items that would otherwise be garbage, such as old candy containers. Any container used should be impervious to solvents, and *not* returned to kitchen duty when you are through!

Have an assortment of large, medium, and small wash containers ready at hand. Some engine parts (like cams and cranks) are large enough that they will not fit into most household ones, but you can at least (carefully!) stand them up in a large container and let it catch the drippings while you clean.

Towels

There are a few different types of towels that you'll want to have around to help wipe parts clean. Shop rags are useful for very greasy situations, as well as for hand wiping. For slightly more delicate situations, find a good quality paper towel with a thick weave, and get at least a 12 pack of them, you'll go through them quickly! Even the best paper towel will leave some lint behind, and because of this, lint-free cloths are recommended for final cleaning operations. These are a little harder to come by (many are advertised as "non-linting" but really aren't), but are worth having. The best are those known as assembly wipes, and they're readily available from engine supply outlets.

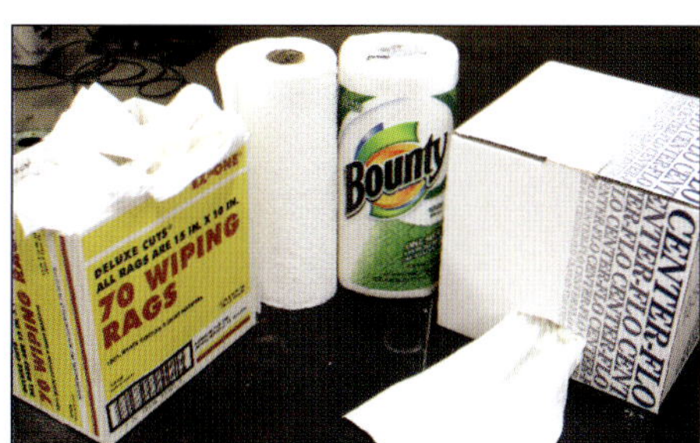

Shop rags and paper towels will do the dirty work of initial parts wiping, and you should have an ample supply of them. Final cleaning should be performed with lint-free cloth like these assembly wipes from Goodson (right). Not pictured here but also helpful are so-called shop towels, which are usually blue and are essentially a stronger type of paper towel.

Engine brush kit

You'll need more than just an old toothbrush to properly clean your engine

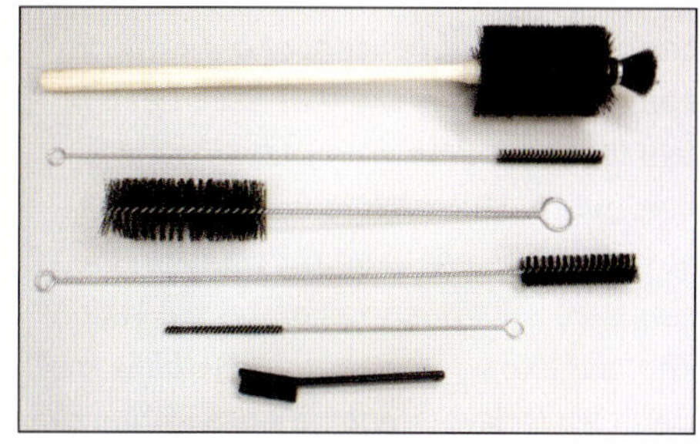

A typical engine brush kit. The nylon bristles won't scratch metal surfaces, and their long handles will enable you to scrub oil galleries spanning the length of the entire block. This particular one lacks a large enough bore brush, though one isn't essential for the job (bores are better cleaned with towels anyway).

parts. Here's one item you won't find at your local hardware store: a dedicated engine brush kit, inclusive of a variety of specialized brushes for cleaning everything from cylinder bores to the smallest crankshaft oil passage. A good one will also have one or two small hand-type brushes for surface cleaning as well. Again, Goodson is a good source for one of these.

Compressed air

It is critical that you have a source of compressed air, as this will allow you to blow contamination out of bolt holes, oil passages, and other areas that cannot be physically reached (or often, even seen). If you don't already have one of these in your garage or shop, small units are inexpensive and will do the job just fine. You'll want to purchase one that can supply at least 80 psi or so.

It's pretty much impossible to thoroughly clean engine parts without an air compressor, but be wary of old compressed air tanks that have been in use for some time. Compressing air naturally pulls moisture out of it, and a tank that has not been regularly drained can have a lot of corrosion on the inside. The last thing you want to do is blow rust particles into your engine parts! A blowing nozzle is mandatory, of course.

Miscellaneous items

Other items you may find helpful during the cleaning process, in no particular order, include: razor blades (to scrape dirty surfaces); a flashlight (to check for

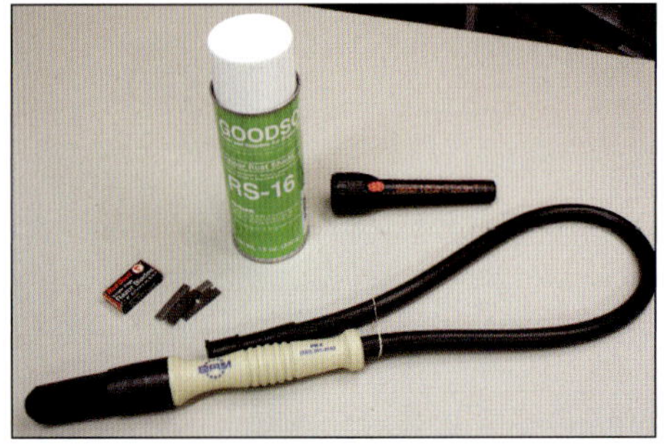

A smattering of various items you may find helpful during cleaning. At the bottom of the photo is a so-called parts washing brush, which you can attach to a specialized spray bottle and pump cleaning solvent through.

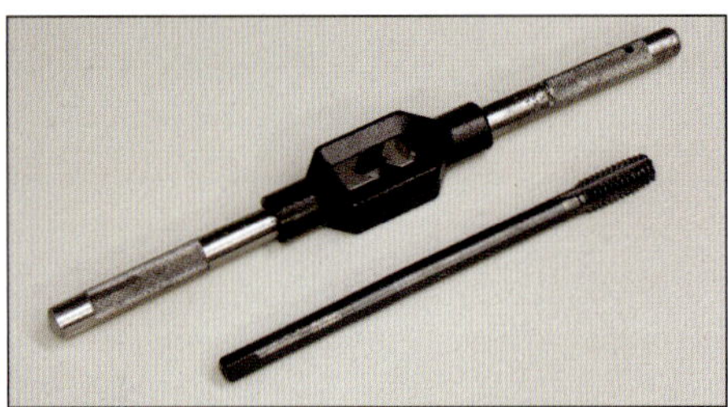

Head bolt holes in particular will have a ton of thread compound left over from the factory. This ARP clean-out tap is designed especially for the M11 head bolt holes in nearly all Gen III or IV LS engines, and can be used with a tap handle from any tap kit. You can buy other size clean-out taps for other holes on your engine block from suppliers like Goodson.

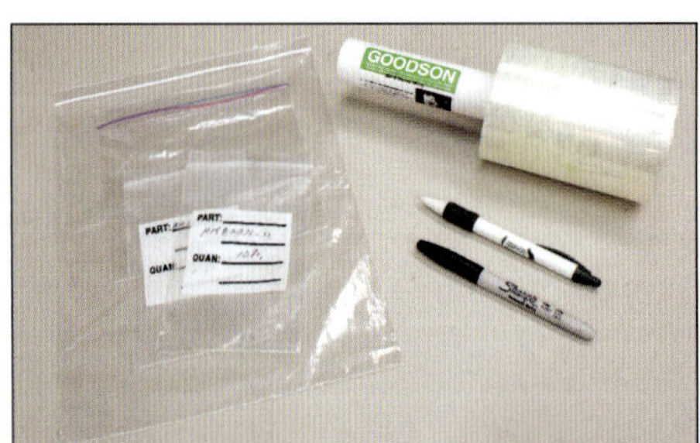

Clean storage bags (and a method of labeling them) will come in handy to put clean parts in and remember what they are during assembly. It should go without saying that you shouldn't reuse the bags you placed your parts in after disassembly in Chapter 3! Parts wrap will be useful for large and oddly shaped items like a crank, cam, or even the engine block.

dirt in dark passages); general washing brushes; thread cleaning taps (to remove stubborn buildup in threaded holes); metal protectant (a dedicated rust inhibitor, or WD-40 if you don't have any); as well as storage bags to cover clean parts.

Some of you may have noticed that, with the exception of its use to rinse off engine degreasers, water and water-based solutions are conspicuously absent from the supply list above. Simply put, we don't recommend washing down any metal parts of an engine with water of any kind. If a component is dirty enough to tempt you to get a hose out, take it back to the machine shop and have it jet washed! The danger, of course, is that water invites corrosion. Though theoretically it is possible to clean parts using water if you immediately dry and spray them with rust inhibitor, the problem is that water may sneak its way into nooks and crannies of a part that you may never see. In this way, it may never evaporate—or may leave damage by the time it does. Keep in mind that even though many parts of Gen III/IV engines are made from aluminum, there still is a lot of iron to rust—especially with an iron block! Again, the *only* use for water that we condone is with rinse-off-type engine degreaser (such as "Gunk"), which should only be used on plastic parts like intake manifolds that do not touch engine oil. *Never* use water to clean internal engine parts!

Safety First!

An engine build is wrought with safety hazards, and a lot of them have to do with dropping heavy parts onto your person. But easily the most dangerous steps of any engine rebuild project are during parts cleaning. The chemicals used can harm the skin and eyes, and their fumes can be toxic, or even flammable. Not taking the proper precautions here could be the biggest mistake of your life.

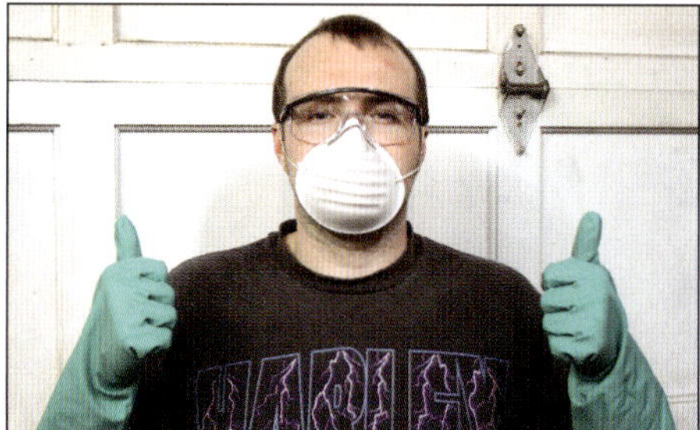

Gloves, safety glasses, and a breathing mask may look lame, but they can save you from health problems. You may see us not wearing gloves in the following photo sequences, but rest assured: your skin will suffer if you choose the same path. The choice is yours.

Be sure to read all warning labels to inform yourself of all chemical hazards, as well as any mixing danger of the solvents you are using. Have an open workspace ready with lots of ventilation. Depending on what you're using to clean, even breathing masks may be a good idea. Latex and/or other types of rubber gloves should be worn at all times, especially when using harsher chemicals. Goggles will be needed to protect your eyes, whether it's from the occasional chemical splash or while using compressed air. These items won't get your engine parts any cleaner, but they're mandatory for your safety!

Final Hints

Because you'll be up-close-and-personal with your engine parts during component cleaning, it's a great time to keep one eye open for any problems like damaged new parts, or refurbished parts that have had machine work incorrectly performed. (In-depth pre-assembly procedures using precision tools will be detailed in the next chapter, although the human eye can spot some problems that no tool can measure.) Should you run across any of these issues, don't lose your cool: remember, the sooner you discover and address them, the better off your rebuild project will be!

Here's a perfect example of a problem that can be visually spotted during component cleaning. This block was cracked even before machining began, and the honing process revealed cylinder wall weakening that slipped the trained eye of the technician (and was not discovered until during the at-home cleaning process). It's better to spot an issue like this now than when your engine is half-assembled!

Step-By-Step Engine Component Cleaning

1 Ready to Begin Cleaning

Before getting started, make sure that the area you'll be working in is cleaned up. The floor should be swept and all table surfaces should be wiped down or covered with plastic. With all of your cleaning supplies close at hand, don some old clothes as well as all of your safety equipment. Set aside a large amount of time to do this–you won't want to take breaks and let contamination set in, or be interrupted and forget what you were doing. You will also want hand cleaning solution nearby to frequently wash up, preventing cross-contamination of parts. This will also help you get any harsh chemicals off of your skin ASAP!

2 Initial Block Inspection

With your block mounted on an engine stand, remove the main caps (if you haven't done so already). Also remove any and all engine block plugs–oil gallery plugs in particular should not have been left in during machining. Begin inspecting all areas of the block for any and all contamination. Too much debris, and you'll want to take the block back to the machine shop for a jet wash. Don't be unreasonable about what you can accomplish here–the removal of a significant amount of dirt or metal shavings requires special equipment you simply do not have.

Important!

3 Inspect and Clean Threaded Block Holes

Inspect all threaded holes in the block, especially main and head bolt holes. Don't assume your machine shop took care of this for you, even if you asked them to! Any significant leftover gunk needs to be cleaned out so that bolts (or studs) can thread in properly. Factory head bolts utilize locking compound on their threads, and their holes normally have more than enough left over to create a problem when the time comes for new fastener installation. Chase any such hole with a clean-out tap, being sure to use lubricant on the threads. Threaded holes can also be brushed if the buildup is minor, but this still isn't as effective as a cleanup tap. If you detect any thread damage anywhere on the block (or on any other engine component for that matter), consult "Fasteners and Threads: Problems and Solutions," on page 23.

Critical Inspection

4 Inspect Block Oil Galleries

Using a flashlight, visually inspect and make note of the following oil passages in the block (see "Oil Galleries" on page 83 to help you pick out which is which):

1. Gallery running from oil pump outlet port to front driver side of block

2. Gallery running from front driver side to rear driver side of block

3. Oil filter passages at bottom rear driver side

4. Main gallery running from driver side rear up diagonally to lifter galleries

5. Passage at rear of block running from lifter galleries upward to oil pressure sensor and/or valley cover. Note that this hole is only threaded on Gen III engines. Gen IV engines used this hole as the oil feed for the LOMA (if equipped) and had the oil pressure sensor thread into the top of the valley cover or LOMA

6. Two lifter galleries, running from rear of block forward to front of block

7. Passages leading from lifter galleries to main bearings (and which intersect cam bearings)

8. Passages leading from lifter bores of cylinders 1, 4, 6, and 7 upward to valley cover or LOMA (some Gen IV engines only)

5 Clean Block Oil Galleries

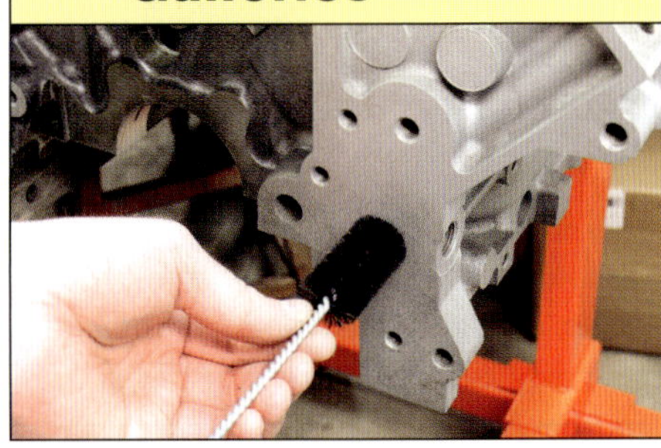

Whether or not you detect visible dirt in any of the oil galleries you've just noted, they must all be cleaned. A good engine brush kit has a variety of engine brushes to choose from. Using the appropriate size brush soaked in mineral spirits, scrub each oil gallery we listed in the preceding step. The main oil gallery is one place that a brush with a very long handle will be needed. A twisting motion helps dislodge dirt as you scrub. You should take the brush out and rinse/re-soak in mineral spirits while scrubbing each passage a few times. After doing this, take the brush and wipe it on a paper towel. If you see any dirt come off, you will need to keep scrubbing. Depending on your style of engine stand, you may have to take the block on and off of it to get better access to the rear of the engine during this step.

Oil Galleries: Delivering the Gen III/IV's Lifeblood

Of all the parts that make up an LS engine, one of the most important to thoroughly clean is the engine block, and nowhere is this more true than with its oil passages. (Whether you choose to call these passages "galleries" or "galleys" is up to you–either nomenclature is considered correct.) A cursory glance at the block may not reveal just where all of these passages are, so it helps to actually visualize the oil flowing through the block, as in the accompanying photo. Look closely and you'll notice white arrows depicting oil flow. As you can see, after exiting the front-mounted oil pump, oil flows sideways to the front driver side of the block, then flows rearward the entire length of the block. The so-called barbell restrictor routes the oil through passages to and from the oil filter, after which the oil flows upward along the rear of the block to the lifter galleries (as well as to the oil pressure sensor or LOMA, as the case may be). The lifter galleries feed passages leading to the cam bearings and main bearings. Many Gen IV engine blocks have additional oil passages running vertically that intersect certain lifter bores.

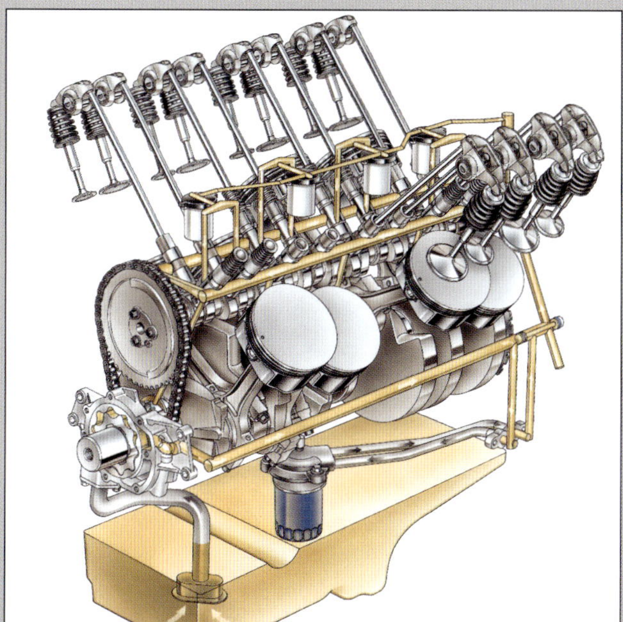

OK, we ran this drawing already in Chapter 1, but it's useful enough that we're going to show it again! This is a schematic of oil flowing through a typical LS engine; while slight differences exist between the oil passages of different Gen III/IV engine blocks (i.e., LSA and LS9 blocks have passages for oil-jet piston cooling), they are all similar. These passages aren't just sealed by plugs: during disassembly, you should have noticed how the rear engine cover seals the rear of the lifter galleries (and also holds the so-called barbell restrictor in place), and how the cam retainer plate seals the front of the lifter galleries. Also note that some of the uppermost oil flow routes shown are part of the LOMA on some Gen IV engines, like this 5.3L LH6. As this particular engine lacks VVT, it also lacks the oil-pressure-fed cam phaser mounted on the cam sprocket (which is fed through an oil gallery running through the camshaft, leading forward from the second journal). (Illustration courtesy of General Motors)

Because these passages would be impossible to accurately cast into the block, they must be drilled at the factory, and it is for this reason that there are a number of plugs at various points along the oil flow path. Again, while you should have taken these plugs out during engine disassembly (and they should also not have been in place during block machining), if for some reason they are in place now, remove them to provide unobstructed access to view and clean the passages.

Here is a typical Gen IV block with vertical pillars running upward from the lifters of four of the cylinders. These pillars have passages that receive oil from the LOMA on AFM-equipped engines. Some engines that were not equipped with AFM had the pillars, but did not have their passages drilled from the factory. But strangely, the passages are present on some Gen IVs without AFM as well, as is the case with this LS2 block!

6 Clean Other Block Passages

Scrub all sixteen lifter bores using the appropriate-size engine brush, as well as all five cam bearing bores. Since you've probably already had your machine shop install cam bearings into these bores, either an extremely soft bristle brush must be used, or you can simply use a mineral spirits-soaked towel to wipe them (our preferred method). The latter technique will require reaching between the block bulkheads for the middle bearing bores. Also brush the crank sensor hole on the lower rear passenger side of the block, the cam sensor hole at the top rear of the block (Gen III engines only), the holes for the main cap side bolts, and any other unthreaded hole you see.

Safety Step

7 Blow Out Block Passages

You should have been wearing them all along, but now is the time to definitely make sure you have safety goggles on. Blow out all passages you have scrubbed with compressed air. Also blow out the block coolant passages, including those leading from the water pump to the block as well as those running between the block and heads. Note that we didn't scrub any of these coolant passages since they are not as critical (and are probably clean enough from the washing done at the machine shop); you can probably just blow them out and visually inspect to make sure nothing is in there.

Important!

8 Clean Cylinder Walls

Equally as important to clean as the oil galleries are the cylinder walls. Again, unless your machine shop did a completely superb job here, there will be leftover particles imbedded into the cylinder walls from the honing process. Don't take the machine shop's word for it that they're ready for assembly! If your engine brush kit included a large bore brush, you should soak it in mineral spirits and scrub each bore. Then, move to paper towels and, moving in multiple directions, scrub with mineral spirits. Acrylic lacquer thinner can also be used, but be careful with this stuff. Final bore cleaning should be done using towels (preferably, non-linting) and automatic transmission fluid. Turning the towel often, continue until every inch of each bore has been scrubbed with ATF and no noticeable particulate is seen on the towel. Take note that you must perform this bore cleaning process one more time after ring fitting, so you don't have to go crazy here if cleaning for pre-assembly.

9 Clean All Other Block Surfaces

Using mineral spirits, wipe every other surface inside the block, as well as all gasket sealing surfaces (for the front and rear covers, etc.); you may use a razor blade to scrape these surfaces (not recommended on deck surfaces), so long as you are careful not to damage the soft metal of aluminum blocks. Watch that your towels do not send lint down any oil passages (or anywhere, for that matter). Here again, you'll need to turn your towel often, add mineral spirits, and rub until little or no dirt is seen on it.

10 Crankshaft Cleaning

The crankshaft is a somewhat bulky item, so to make things easier and safer, it's best to mount it on an inspection stand during cleaning. (If you don't have one of these, you can also bolt a flywheel or flexplate on the back and have the crank standing vertically while you clean.) Clean all journals and other surfaces using non-linting towels with mineral spirits, and similarly brush the small oil passage holes in each journal. Also, on most factory and aftermarket cranks, there is a hollow hole that runs the length of the crankshaft down its centerline; now is a good time to look and make sure its rear plug is in place, or you're in for a major oil leak. (Consult your machine shop if the plug is not there.)

11 Clean Other Large Internal Engine Parts

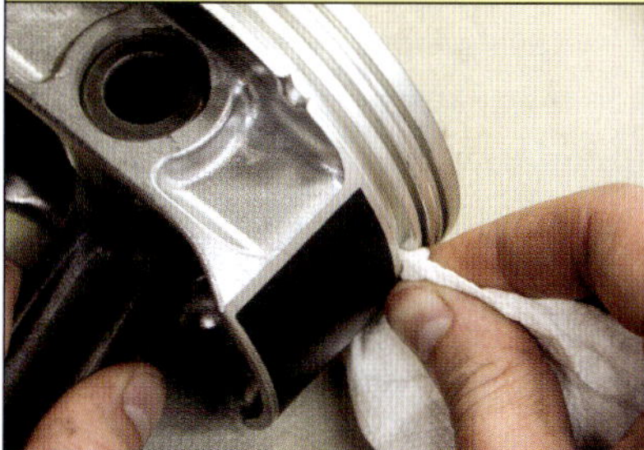

Clean all surfaces of the cam, pistons, and connecting rods using clean non-linting towels and mineral spirits. Be mindful of the piston ring grooves. Also, Gen III and IV cams are hollow (some not all the way through), so you should brush the inside of the cam, and in addition, the oil passages in the second cam journal from the front (this last point applies to VVT-equipped engines only). Have a look at the bolt holes in the connecting rods and front of the cam for any contamination as well.

12 Solvent Soak Small Parts

Many smaller internal components—such as pushrods, bearings, and piston rings—can be fully soaked for the ultimate in cleanliness. This also works well with any bolts you may be reusing. Let the parts sit in a bath of clean mineral spirits awhile, then remove and wipe them off. Don't forget to blow out any internal passages, especially with pushrods. Note: if you choose to clean piston rings this way, do not mix the compression rings up if you've already file-fit them and are cleaning for final assembly!

13 Non-Soakable Parts

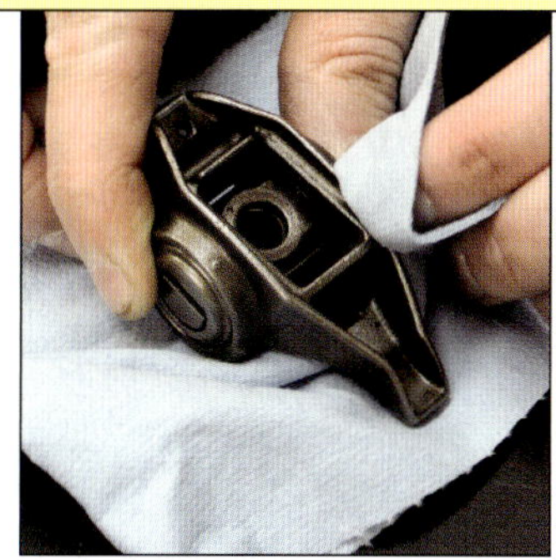

Solvent soaking lifters and rocker arms is not recommended. You don't want to pull the oil out of the inside of these items and destroy them at initial startup—their internal needle bearings need the lubrication. An external mineral spirits wipe will suffice. If these items are particularly dirty (for example, if they are being reused, this is strongly discouraged for lifters) and you insist on soaking them, you'll need to subsequently submerse them in oil for a long period of time and move them under the surface to work the lubrication in. The same thing goes for your timing chain.

14 Cylinder Head Cleaning

If you're reusing your stock cylinder heads, you should have had your machine shop wash them after performing the required machine work. If they did not do this, take your heads back and have them do it for you before performing your own cleaning. Brand-new cylinder heads require their fair share of cleaning, too. If your heads are not currently assembled, scrub the valve guides and other holes using brushes soaked in mineral spirits. Blow out all pushrod holes, bolt holes, intake/exhaust ports, and other passages in the head using compressed air (safety glasses, folks!). Ensure that the exhaust manifold bolt holes are clear of threadlocking compound (used heads only). Finally, wipe all head surfaces with mineral spirits.

16 Clean Intake Manifold

Your intake manifold may be dirty on the outside, but this pales in comparison to the state of affairs on the inside, where there will at a minimum be a coating of oil (thanks to your PCV system). The intake is cleaned most easily using a spray-on, rinse-off engine degreaser (water is OK to use here because there is no worry of corrosion with plastic). Before doing this, remove any sensors as well as the eight factory fluorosilicone intake port gaskets (car intakes only; truck intakes use carrier-style gaskets) and throttle body gasket. Use new ones; discard the old ones. Make sure you get all gasket grooves clean, or you risk compromising gasket seal.

15 Cleaning Engine Covers

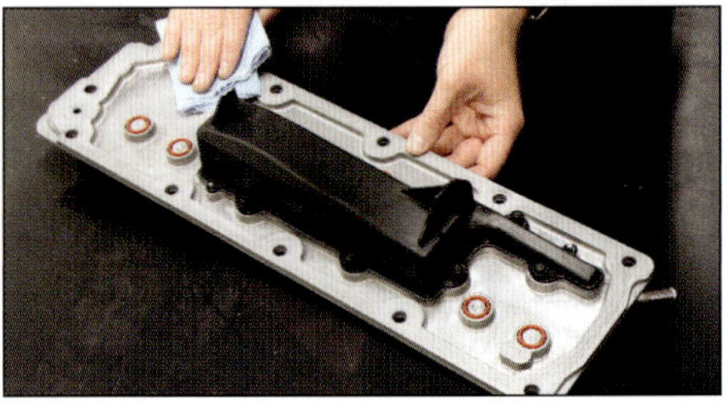

Clean the engine covers, including the front cover, rear cover, valley cover/LOMA assembly (as applicable), and oil pan. Pay attention to any residue on the gasket sealing surfaces. Special care must be taken with the oil pan, which will be particularly gummed up. Different pans also have varying baffling setups in them, as well as different fittings and a bypass valve about the oil filter, so be sure to remove any such items (and replace with new items as necessary) during cleaning so you can get into every nook and cranny. Some of these engine covers have sensors like cam position sensors, temperature sensors, or oil level sensors that you'll want to be careful of, and the LOMA will be particularly delicate, what with electrical connectors and lots of seals and solenoids. Any under-valley-cover baffles (for the PCV system, as with the valley cover shown) should be removed and the inside cleaned, or at a minimum flushed with solvent and blown out. Also note the condition of O-ring-style seals on the underside of Gen IV valley covers. Do not expose these to harsh solvents, and replace them if they are not in good condition– these things seal oil galleries! The knock sensor grommets on the underside of Gen III valley covers should be wiped as well (or if in poor condition, removed and replaced).

Should I Replace My LOMA?

In the relative scheme of things, GM's cylinder deactivation technology, with its special switching lifters and Lifter Oil Manifold Assembly, hasn't been around all that long. Because of this, the jury is still out on whether and when the LOMA used on engines so equipped will need to be replaced. If you're rebuilding an AFM-equipped engine, you should definitely replace its special switching valve lifters with new ones (along with the rest of your lifters of course), and you'll also need to decide whether to spend the money on a new LOMA (or replacement components thereof).

As no firm recommendations exist from GM as of the time of print, one thing that we can say is that because this item contains solenoids, it's a safe bet that their life is limited in some fashion. If your engine has a LOMA, a rebuild is an opportune time to replace it, or at least its solenoids. It's better to do this now than regret it later!

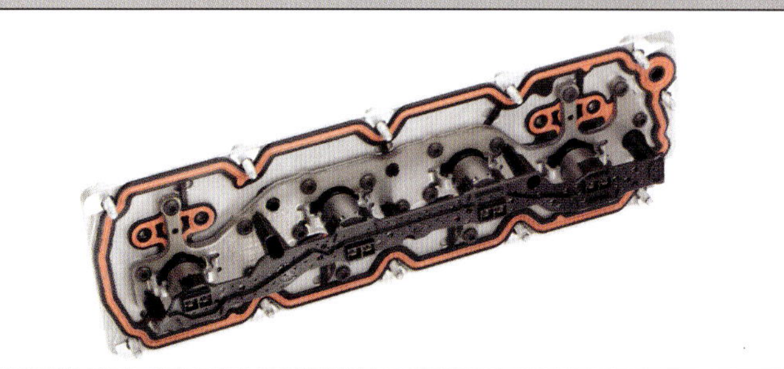

The Lifter Oil Manifold Assembly, or LOMA, looks like this and is used in place of an ordinary valley cover on 2005 and later Gen IV engines equipped with Active Fuel Management (a.k.a. Displacement-On-Demand). As you can see, it integrates solenoids into the valley cover, and these direct pressurized oil flow to special ports on eight of the engine's lifters to deactivate them under light engine load. (Photo courtesy of General Motors)

17 Clean Throttle Body

Clean your throttle body, which more than likely will have some carbon buildup inside. Carb cleaner works best for this, but make sure it's the sensor-safe stuff (use so-called "intake/throttle body cleaner" if you're not sure). Remove all sensors from the throttle body before doing so; it's a good idea to acquire new ones of these if your engine had a lot of miles on it. Take special care with the electronics on drive-by-wire throttle bodies. Here again for engines that saw a lot of miles, replacing such a throttle body's motorized actuator may be a good idea (shown intact and removed side-by-side).

18 Clean Other EFI-Related Hardware

For best performance, used fuel injectors should be professionally cleaned and flow-tested, as they can clog over time from impurities in gasoline or E85. If there is an incurable problem with the injectors, they can be replaced. As far as what you can do yourself, you'll at least want to remove and discard the O-rings at the top and bottom of the fuel injector and acquire new ones. Fuel rails should, at a minimum, be sprayed out with sensor- and fuel-injector-safe carb cleaner, with particular attention being paid to injector O-ring seating surfaces for damage.

Don't contaminate your engine by using dirty tools. Just think of all of the rusty, greasy bolts you've worked on in the past with your socket wrenches, and you'll understand the kinds of small particles that can reside on their surfaces and in nooks and crannies. This is important to keep in mind not only during pre-assembly, but during final assembly as well.

Clean engine oil

Pre-assembly requires tightening many fasteners to their final specifications. Though many are designed to be installed "dry," some fasteners require lubrication to help achieve an accurate torque reading. Engine oil is the most commonly used thread lubricant, though some aftermarket fasteners require specialized moly-based thread lubricant, which is often included. Because moly lubricant is often supplied in small amounts, it should be saved for final assembly and the alternate "with oil" torque spec used during pre-assembly (unless no such specification is provided). Also, you will need to physically lubricate engine bearings to allow spinning the crankshaft during pre-assembly. A good set of small brushes and/or an oiling can is less messy and more sanitary than applying lube with your fingers.

Parts cleaning supplies

You're already quite familiar with the cleaning supplies used for engine parts, and it's good to have at least some

Quality non-synthetic engine oil will aid in lubricating bearings and other surfaces during pre-assembly. Heavier-weight oils such as SAE 30 are preferred as they stick to surfaces better than lightweight oils like 5W-30. Also, if you have special assembly lubricants for bearings, now is probably not the time to use them, unless you'll have a sufficient amount left over for final assembly.

Step-By-Step Pre-assembly Procedures

1 Ready to Begin Pre-assembly

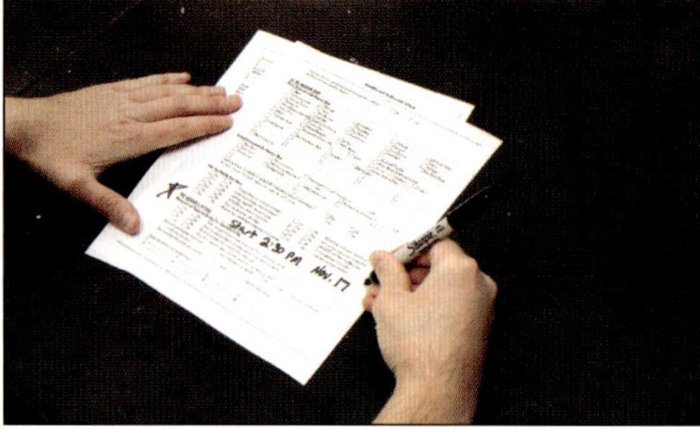

Pre-assembly is easily the most time-consuming part of any engine build, so there is almost no way you will get it done in one sitting. It's OK to stop and do something else for a while—just be sure that you take careful notes, and do your best to avoid immediate emergencies that might arise and prevent you from leaving your work at a natural stopping point. Have a copy of our Work-A-Long sheet (available at cartechbooks.com) ready and be sure to check off each item as you go, and have temporary means of keeping items clean (storage bags, etc.) while you are away.

Precision Measurement

2 Measure Engine Block Machined Surfaces

Chances are, your home measuring tools aren't as accurate as those of a machine shop, but this doesn't mean you can't use them to double-check whether your engine block machine work was performed correctly. Utmost trust in your machine shop may tempt you to skip this step, but before doing so, keep in mind that machinists can make mistakes, too! Using the tightening specifications (sequence optional) shown in Steps 7-8 in Chapter 8, install each main cap and use a dial bore gauge to measure the diameter of each main bearing bore in a crosswise pattern (minimum two measurements per bore). All measurements should be consistent within 0.001 inch, with a maximum out-of-round of less than half a thousandth on any given bore. Remove the main caps and perform the same procedure on each cylinder bore, measuring the bore in a few different places along its height. Record these measurements, they should all be within 0.001 inch as well.

Should I Replace My LOMA?

In the relative scheme of things, GM's cylinder deactivation technology, with its special switching lifters and Lifter Oil Manifold Assembly, hasn't been around all that long. Because of this, the jury is still out on whether and when the LOMA used on engines so equipped will need to be replaced. If you're rebuilding an AFM-equipped engine, you should definitely replace its special switching valve lifters with new ones (along with the rest of your lifters of course), and you'll also need to decide whether to spend the money on a new LOMA (or replacement components thereof).

As no firm recommendations exist from GM as of the time of print, one thing that we can say is that because this item contains solenoids, it's a safe bet that their life is limited in some fashion. If your engine has a LOMA, a rebuild is an opportune time to replace it, or at least its solenoids. It's better to do this now than regret it later!

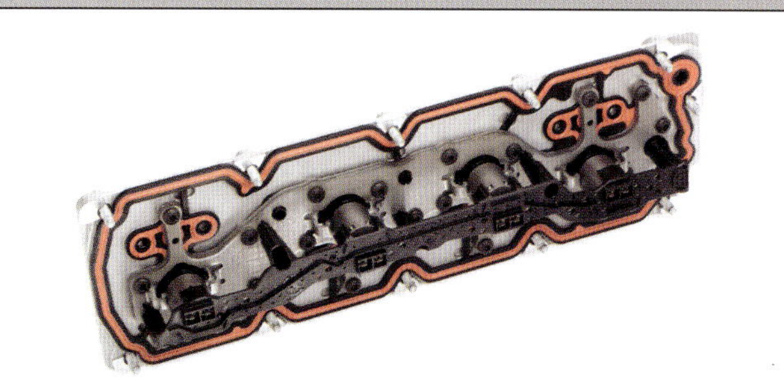

The Lifter Oil Manifold Assembly, or LOMA, looks like this and is used in place of an ordinary valley cover on 2005 and later Gen IV engines equipped with Active Fuel Management (a.k.a. Displacement-On-Demand). As you can see, it integrates solenoids into the valley cover, and these direct pressurized oil flow to special ports on eight of the engine's lifters to deactivate them under light engine load. (Photo courtesy of General Motors)

17 Clean Throttle Body

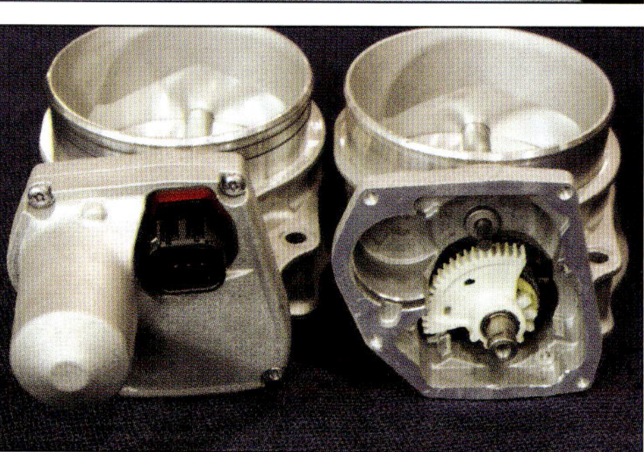

Clean your throttle body, which more than likely will have some carbon buildup inside. Carb cleaner works best for this, but make sure it's the sensor-safe stuff (use so-called "intake/throttle body cleaner" if you're not sure). Remove all sensors from the throttle body before doing so; it's a good idea to acquire new ones of these if your engine had a lot of miles on it. Take special care with the electronics on drive-by-wire throttle bodies. Here again for engines that saw a lot of miles, replacing such a throttle body's motorized actuator may be a good idea (shown intact and removed side-by-side).

18 Clean Other EFI-Related Hardware

For best performance, used fuel injectors should be professionally cleaned and flow-tested, as they can clog over time from impurities in gasoline or E85. If there is an incurable problem with the injectors, they can be replaced. As far as what you can do yourself, you'll at least want to remove and discard the O-rings at the top and bottom of the fuel injector and acquire new ones. Fuel rails should, at a minimum, be sprayed out with sensor- and fuel-injector-safe carb cleaner, with particular attention being paid to injector O-ring seating surfaces for damage.

19 Clean All Other Miscellaneous Parts

Clean all other parts we haven't mentioned specifically, such as the crankshaft oil deflector, main bearing caps, and oil pump pickup tube (make sure its screen is completely free of debris). On a similar note, you will want to keep certain solvents away from rubber and plastic, so use care when cleaning the "barbell restrictor" and any O-rings. A dry cloth wipe is the best way to clean cam, crank, and other sensors (if they came from an engine with a lot of miles, most of these items should be replaced anyhow for optimum reliability). Items like the cam retainer plate need to be treated with care as well (there is a built-in seal on the back). Inspect all parts before you clean, and definitely before you soak!

Important!

20 Prep Items for Temporary Storage

It's best to begin your engine pre-assembly or final assembly as soon as possible after completion of parts cleaning. If the parts will need to be stored for any length of time, however, you will want to take steps to ward off rusting of any iron or steel parts. Cylinder walls and crankshaft journals are particularly susceptible to corrosion, and it's advisable to spray those items immediately with a protectant like WD-40 or, better yet, a specialized rust inhibitor like this one from Goodson. All parts of an iron block should be cared for as well if it will be sitting awhile (by the way, now is an excellent time to mask off and paint the outside).

21 Store Clean Items

If you're not pre-assembling or assembling immediately, all cleaned items should be stored in plastic bags or with specialized wrap like this stuff from Goodson. Less critical items like bolts, engine covers, and sensors are OK in a clean cardboard box or sealed plastic container. A large garbage bag works well for larger items like engine blocks. Make sure any solvent has evaporated before sealing anything off. Congratulations! You've completed component cleaning and can now move on to pre-assembly (or final assembly, as the case may be).

CHAPTER 7

Pre-Assembly Procedures

Let's be honest: at this point, you're probably getting a bit tired of reading each chapter introduction. Each one contains parent-like nagging of the importance of the material therein, and threatens you with grave consequences for not paying attention to detail. However, attention to detail is what an engine rebuild is all about, and even though you've successfully come this far, there are still many important steps to undertake prior to final engine assembly. It is this critical time that this chapter is all about.

This phase is best known as pre-assembly, and now is when you'll determine once and for all whether the parts that comprise your engine will actually work together. This is also the time when you'll be performing checks to make sure all machine work was done properly to the engine block and other components. If there is one piece of advice that you must understand right now, it is this: *Do not assume that your parts will fit together, even if you have been assured they will work properly!* It's a lot easier to partially assemble your engine now and perform the required checks, rather than discover part fitment problems during final assembly. Worse yet, failure to pre-assemble could leave outstanding issues that might not even be found during final assembly; we shouldn't have to tell you that damage resulting from physical parts interference, improper bearing clearances, or other problems in a running engine can be swift, severe, and oh-so expensive!

Do I Really Need to Pre-assemble?

The short answer to whether every engine needs to be pre-assembled is a resounding "yes!" But this blanket statement is a little deceptive, as in reality, the fact that no two LS engines are alike means required pre-assembly is a matter of degree. All engine builds—even stock rebuilds—will require at least some amount of pre-assembly, mainly to establish basic items like proper bearing clearances and piston ring end gap (this includes verification for "drop-in" rings, mind you!). On the other hand, stock rebuilds will normally have no need to undergo more advanced checks like rotating assembly clearances or degreeing of a stock cam; pre-assembly procedures like these are far more important and involved for high-performance engines, as the likes of bigger cams and longer-throw crankshafts all mean special care and attention must be paid (and often, a few modifications made!).

Because of the discrepancies in pre-assembly necessities, we'll mention which procedures apply to all rebuilds—and which are for high-performance engines only—as we undergo the step-by-step pre-assembly process below. We suggest that you read through all steps *first* to get a better idea of which will be needed for your engine. This will fill you in on exactly what tools you will need to have handy, as well as any special considerations you'll want to know, *before* beginning.

Pre-assembly Supplies

While all tools required for pre-assembly were listed in Chapter 2, here are the miscellaneous supplies you'll need during the process, along with a few final suggestions.

Clean area to work in

You just cleaned all of your parts using the procedures in Chapter 6, and know that you'll need to go through the cleaning process one more time before final assembly. To make your life a little easier, you should do your best to keep all parts as clean as possible during the pre-assembly process, and this all starts with having clean worktables, tools, and a dust-free environment. Though components will no doubt get contaminated during pre-assembly, why not minimize this instead of setting yourself up for even more cleaning work?

Don't contaminate your engine by using dirty tools. Just think of all of the rusty, greasy bolts you've worked on in the past with your socket wrenches, and you'll understand the kinds of small particles that can reside on their surfaces and in nooks and crannies. This is important to keep in mind not only during pre-assembly, but during final assembly as well.

Clean engine oil

Pre-assembly requires tightening many fasteners to their final specifications. Though many are designed to be installed "dry," some fasteners require lubrication to help achieve an accurate torque reading. Engine oil is the most commonly used thread lubricant, though some aftermarket fasteners require specialized moly-based thread lubricant, which is often included. Because moly lubricant is often supplied in small amounts, it should be saved for final assembly and the alternate "with oil" torque spec used during pre-assembly (unless no such specification is provided). Also, you will need to physically lubricate engine bearings to allow spinning the crankshaft during pre-assembly. A good set of small brushes and/or an oiling can is less messy and more sanitary than applying lube with your fingers.

Quality non-synthetic engine oil will aid in lubricating bearings and other surfaces during pre-assembly. Heavier-weight oils such as SAE 30 are preferred as they stick to surfaces better than lightweight oils like 5W-30. Also, if you have special assembly lubricants for bearings, now is probably not the time to use them, unless you'll have a sufficient amount left over for final assembly.

Parts cleaning supplies

You're already quite familiar with the cleaning supplies used for engine parts, and it's good to have at least some

Step-By-Step Pre-assembly Procedures

1 Ready to Begin Pre-assembly

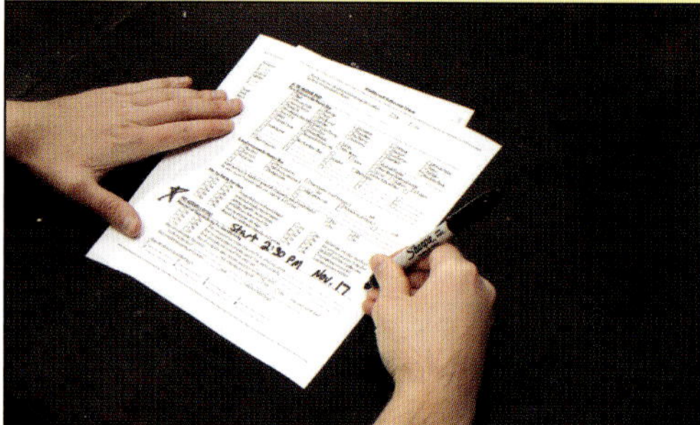

Pre-assembly is easily the most time-consuming part of any engine build, so there is almost no way you will get it done in one sitting. It's OK to stop and do something else for a while—just be sure that you take careful notes, and do your best to avoid immediate emergencies that might arise and prevent you from leaving your work at a natural stopping point. Have a copy of our Work-A-Long sheet (available at cartechbooks.com) ready and be sure to check off each item as you go, and have temporary means of keeping items clean (storage bags, etc.) while you are away.

Precision Measurement

2 Measure Engine Block Machined Surfaces

Chances are, your home measuring tools aren't as accurate as those of a machine shop, but this doesn't mean you can't use them to double-check whether your engine block machine work was performed correctly. Utmost trust in your machine shop may tempt you to skip this step, but before doing so, keep in mind that machinists can make mistakes, too! Using the tightening specifications (sequence optional) shown in Steps 7-8 in Chapter 8, install each main cap and use a dial bore gauge to measure the diameter of each main bearing bore in a crosswise pattern (minimum two measurements per bore). All measurements should be consistent within 0.001 inch, with a maximum out-of-round of less than half a thousandth on any given bore. Remove the main caps and perform the same procedure on each cylinder bore, measuring the bore in a few different places along its height. Record these measurements, they should all be within 0.001 inch as well.

of them nearby during pre-assembly. For example, mineral spirits and clean rags will allow you to quickly deal with accidental spills or contamination before it spreads (or before you slip and hurt yourself!). Also, items like main bearings will need to be installed, cleaned, and reinstalled throughout the pre-assembly process, so be prepared with what you'll need to do this.

What If Anything Goes Wrong?

Much of pre-assembly is geared toward not only finding parts fitment problems, but addressing them. If as you go along you note an issue whose solution is not described in the below steps, the best thing to do is to call your machine shop. Whether it's a potential problem with machine work or difficulty with parts fitment that you can't quite figure out on your own, they're an excellent resource and will often know exactly what to do without even seeing the parts themselves. Sometimes, a part problem you think is unfixable is far from it, and may only require minimal at-home work to solve (or perhaps a quick trip to the machine shop to have them take care of it for you).

That said, let's pre-assemble!

Precision Measurement

3 Measure Other Engine Parts

It will be helpful to check the roundness of other critical surfaces, including but not limited to crank journal diameters (inspect for taper and out-of-round) and connecting rod bearing bores. This is especially true if your machine shop didn't have much of a hand in preparing these (i.e., they are new aftermarket parts). In conjunction with the cylinder bore diameters you've just written down, measuring your piston diameters will allow you to verify piston-to-wall (a.k.a. piston-to-bore) clearance. Proper clearance ranges from less than a thousandth of an inch for some cast pistons and as high as 0.010-inch for forged pistons, though exact spec varies by piston alloy (aftermarket pistons normally include a spec sheet with the recommended clearance listed) as well as the type of clearances your machine shop has had success with in past engines. Error compounding for imprecise measuring tools means you may not get very accurate readings for this, but way-out-of-whack measurements should have you on the phone with your machine shop. As a side note, you may also measure specs on your cylinder heads like valve-to-guide clearance, but your tools probably lack the precision needed for these checks, too.

4 Main Bearing Clearances - Install Main Bearings

With the main caps removed, install your upper main bearing shells in the block. These "upper halves" have holes in them for oil to enter from the block. Note that the center main bearing is the thrust bearing on LS engines. Also put the lower halves of the main bearings in the main caps. Use no lubricant whatsoever on any of these bearing shells–it will interfere with bearing clearance measurement. Once installed, wipe them with a clean cloth to remove any possible contamination.

5 Main Bearing Clearances - Install Crankshaft and Plastigage

Now after giving the crank main journals a quick wipe, gently lay the crank in place in the block. Again, do not use oil on any surfaces–they must be clean and dry. Lay a piece of Plastigage on top of each main bearing journal once the crankshaft is in.

Bearing Clearance Measurement

You saw this item in Chapter 2, and here it is close up: gauging plastic, a.k.a. Plastigage. Without this innocent-looking tiny plastic strip, measurement of bearing clearances is practically impossible!

The term "bearing clearance" refers to the amount of space between the surface of the fast-spinning crankshaft journal and the main (or rod) bearing while the engine is running. Simply put, this space must remain occupied by a certain amount of engine oil to ensure proper lubrication. Too much clearance and you lose oil pressure; too little, and oil can't flow in and out properly. While it's possible to assess bearing clearance by carefully measuring and comparing the bearing bore diameter, the bearing thickness, and the crank journal diameter, the problem is that all but a well-equipped machine shop lacks the kind of super-expensive, precise measuring equipment needed to do this. Even when used correctly, the best home tools are likely only accurate to about half of a thousandth of an inch (which may be just fine if assessing roundness of a crank journal); but for each item you're trying to measure and compare, the error only compounds. Therefore, it's best to go with the tried-and-true method of the home hot-rodder and pro-engine-builder alike: gauging plastic, better known as Plastigage.

Plastigage is made in a few different types designed to measure a specific range of bearing clearances. Most LS builds will need the type that measures clearances in the range of 1 to 3 thousandths of an inch, which is normally green in color. This stuff is cheap, so pick up at least a couple of strips for your project. It's available by mail order or at any machine shop.

Here's how it works. The Plastigage strip gets squeezed between the bearing and the crankshaft journal when the main or rod cap is tightened down, and stays that way when the cap is removed. By measuring the width of the smooshed piece, the bearing clearance is found: more squish = less clearance. It's simple, but Plastigage is quite accurate and has been used in engine assembly for decades.

A few notes on Plastigage usage techniques. Plastigage should not be pulled apart, but rather cut with a sharp knife, preferably a razor blade (shown). Always use one piece of Plastigage across the entire width of the bearing or journal, and make sure it is parallel with the crank centerline. Also, never rotate the crankshaft while the Plastigage is being compressed; this will destroy a proper reading.

We've detailed the procedures for checking both main and rod bearing clearances in the photo captions. However, a few notes on the results you may come up with are in order. A flattened Plastigage piece that varies in width across its length may indicate a problem with crankshaft journal taper, but it could also indicate a bearing shell that wasn't seated properly. Also, be wary of readings that vary between journals, as this can indicate a number of possible issues. These include one or more journals being out-of-round, or (in regard to main journals) a crankshaft with excessive runout. See step 8 for checking runout: cranks that exhibit within-spec (but nonzero) runout can still have bearing clearances checked properly by simply tightening and checking one main bearing cap at a time.

If these types of issues are ruled out and you still have bearing clearances that are either too large or too small, you have options. Though one would be re-machining of the crankshaft, it's far easier to just buy a different set of bearings. If you're looking to tighten up bearing clearances, you'll want to purchase a set of undersized (smaller I.D.) bearings–as mentioned in Chapter 5, these will especially be needed with cranks that have been ground down to a smaller journal diameter. Conversely, loosening bearing clearances would involve getting a hold of bearings with a larger I.D. Both types of sets are readily available for LS engines. Individual oversized or undersized bearing shells can also be used to correct issues like one crank journal having been turned down and the others being of stock diameter (in which case you'd also need to be very careful to use the correct bearing shell in that exact location during final assembly!).

Bearing Clearance Measurement
CONTINUED

Suggested LS Bearing Clearances (in inches)

	Main Bearings	Rod Bearings
Stock to moderate high-performance	0.00080–0.00210	0.00090–0.00250
Moderate high-performance to racing	0.00250–0.00275	0.00225–0.00250

Note: Choice of bearing clearances is one of the more subjective areas of any engine build, so the above should only be considered general guidelines. Most applications–many high-performance engines included–will want to use the stock clearance specifications, which are compatible with lighter-weight oils such as 5W-30 (and will therefore yield slightly improved power and fuel economy versus use of a thicker-weight oil). See your GM service manual for full bearing clearance specifications for your engine. Engines destined for higher-stress applications where higher oil temperatures and extreme RPM will be seen will wish to use thicker oils such as 10W-40 and must increase bearing clearances accordingly, sometimes beyond the suggestions shown in this chart. Bearing clearance and intended oil type are closely intertwined, must be decided on simultaneously, and are highly application-dependent. You should always follow your crankshaft and/or bearing manufacturer's recommendations for clearances, particularly when using an aftermarket crank.

Torque Fasteners

6 Main Bearing Clearances - Install Main Caps

Install the main caps (again, with dry bearings) and tighten them in the sequence shown in steps 7-8 in Chapter 8. However, do not use a hammer to align the thrust bearing surfaces. Also, although their installation will likely not affect your main bearing clearance reading, you should install your main cap side bolts and tighten to spec. (If you purchased new GM side bolts, they will have sealant pre-applied, so don't use them until final assembly. Use the old bolts for now.)

7 Main Bearing Clearances - Remove Main Caps and Measure Plastigage

Loosen and remove all main cap bolts followed by the main caps themselves. Note that the tight fit between the caps and oil pan rails can make removal tricky. As during disassembly, we suggest using two long 3/8-inch extensions to gently pull them upward (rather than buying the special slide hammer tool). Underneath each cap you'll find that the Plastigage has now been smooshed. Match its width to the nearest marking on the Plastigage sheath, and you've got your bearing clearance. See "Bearing Clearance Measurement" for suggestions on proper main bearing clearances and what to do if they are not satisfactory. When you're done, clean all remnants of Plastigage from the crank journals and remove the crankshaft from the engine block.

Precision Measurement

8 Check Crankshaft Runout

Remove all upper main bearing shells from the block except the #1 and #5 (front and rear) shells. Clean these remaining shells, lubricate them with oil, then set the crank in place atop them (lube the #1 and #5 main journals on it, too). Set a dial indicator to measure along the surface of the #3 (center) main journal, being sure the indicator's tip won't drop into the crank oiling hole(s) as the crank is turned. Spin the crank over and note the highest and lowest readings on the dial indicator. The difference between them is what is important; anything more than a thousandth or so is cause for possible concern. Total acceptable runout varies by application, so if you see anything that perturbs you, consult your machine shop for recommendations. They may even have suggestions for straightening your crank that you may find surprising (like torquing your main caps in place and letting this straighten the crank out for you!). You should also move the dial indicator to the rear of the crank and measure runout on the rear flange, where the rear seal rides; runout here can't exceed 0.002-inch. Please note that because there will be no thrust bearing in place during the runout checking process, you may hear the reluctor ring dragging lightly along the rear of the engine block—this is normal.

9 Install Crankshaft with Lubricated Bearings

Now remove the crankshaft. Install and lubricate the remaining upper main bearing shells in the block. Set the crank atop them, lube all main journals, then install all main caps with lubricated lower main bearing shells. Use the instructions and sequences detailed in steps 6 through 8, Chapter 8, for main cap installation and tightening, being sure to use a hammer to align the thrust bearing surfaces! The crank should now move freely by hand. If it does not, repeat the thrust bearing alignment procedure. If this does not cure the problem, something may be wrong with your block's main bearing bores and/or their alignment (possibly requiring line honing).

Precision Measurement

10 Check Crankshaft Endplay

There are a couple of ways to check that your thrust bearing (center, or #3, main bearing) is allowing the proper crankshaft endplay. The simpler way is via a feeler gauge. Insert a large flathead screwdriver between a crank counterweight and a main cap and pry rearward and forward, measuring the space between the thrust surfaces on the crank and the thrust bearing. Measure both the space in front and space behind for comparison purposes; GM recommends between 0.0015 and 0.0078 inch of clearance (always follow the recommendations of your crankshaft and/or bearing manufacturer when using aftermarket components). The more accurate way to measure this is to mount a dial indicator to read off of the crank snout (or rear crank flange) and record dial indicator movements as you pry fore and aft.

Rotating Assembly Clearance Issues: It's a Stroker Thing

Pre-assembly step 20 involves checking for physical interference between parts of the rotating assembly and between the rotating assembly and engine block. These problems will occur almost exclusively in engines using a crankshaft with longer throws than stock, though they can also occur when using certain aftermarket connecting rods (such as those made from aluminum).

The most common stroker interference issue is that the heads of the rod bolts come close to the block near the bottom of the cylinder bore. It is best to have both rods on a journal when checking this, as it prevents side-to-side rod movement that might show a false clearance reading (you can also use only one rod and simply push it toward the outer edge of the journal–clearance is at a minimum here). To obtain adequate room, the best thing to do is to remove material from the block as discussed in step 20. An alternative, however, would be to "nick" the rod bolt heads. This would require: 1) that you have each rod in its final location, and 2) that each rod bolt is tightened to spec using the exact method that will be used in final assembly. This latter method is also not recommended as it can affect engine balance if overdone.

In addition, you should know that extreme-stroke engines can run into interference problems between the connecting rods and the camshaft. This was a substantial problem with earlier small-blocks, but it is very rare with the Gen III and IV thanks to their higher cam position. Such interference would probably not occur unless a very long stroke (well over 4.125 inches) were combined with a bulky connecting rod. As it would be very difficult to fit a stroke of this size in any production block, you wouldn't have to worry about this issue unless building a very large cubic-inch engine based off of an aftermarket engine block. In this case, you'd want to degree your cam (steps 21-25) before checking rod-to-cam clearance.

Cam Degreeing: Verification vs. Alteration

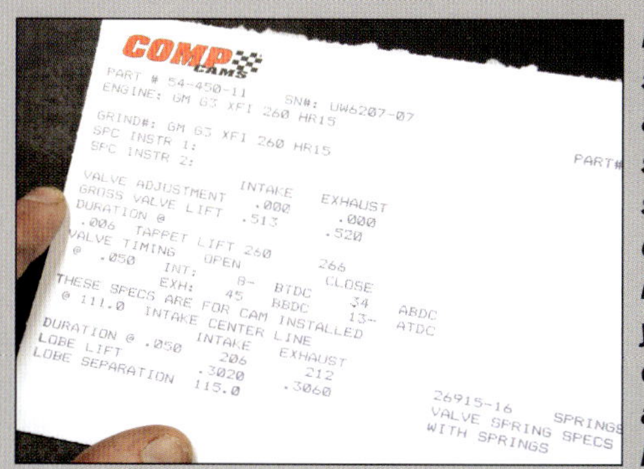

Manufacturer specifications are normally shown in a so-called cam card included with your camshaft. In addition to listing lift and duration, the crankshaft angular positions at which valve events occur are also given. If you have a custom application for a specific cam, you may wish to second-guess the cam designers' choices and have valve events occur sooner than the recommended crank positions listed (or in the alternative, delay them). "Advancing" or "retarding" the cam in this manner can alter characteristics of the engine's power and torque curves: advancing increases low- and mid-range torque at the expense of top-end power, while retarding will have the opposite effect. (LS engines equipped with VVT do this on the fly, yielding the best of both worlds!)

While the term "cam degreeing" often refers to the actual alteration of cam phase with respect to crank position (and the resultant fine-tuning of the engine's powerband), it also can refer simply to the verification of valve event timing as compared with a cam manufacturer's specifications.

Regardless of whether alteration or simple verification is being done, high-performance applications using camshafts with large amounts of lift and duration must go through the degreeing process right now. Since the phase of the cam determines exactly when (and how much) the valves are open when the piston is at or near Top Dead Center (TDC), verifying proper piston-to-valve clearance is impossible without first degreeing the cam. This way, any necessary changes can be made to the valvetrain and/or piston if this clearance is insufficient, alterations that could be far more difficult and costly if their need were discovered later on!

So how to go about alteration? Although factory and stock-replacement timing sets only allow the camshaft to be installed one way–making adjustment of the cam's angular position with respect to the crank impossible–many aftermarket sets provide sprockets with multiple keyways and markings to enable such adjustment (you can see one of these in the last photo of step 25). Still, the most common cam installation practice is known

Cam Degreeing: Verification vs. Alteration *CONTINUED*

as installing the cam "straight up," i.e., using the standard position markings on the crank and cam sprockets. Normally, manufacturers design their cams to be installed just this way, and for most applications, it should yield the proper timing of valve opening and closing events with respect to the position of the piston in the cylinder.

Even if no such alteration is desired, it is still advisable to double-check that the timing of all valve events match up to the specs provided on the cam card (i.e., go thorough the verification process). This ensures that all parts have been manufactured correctly, and we're not just referring to the cam lobes. For example, while the marks on the cam and crank sprockets may indicate TDC for the #1 cylinder, the trueness of these marks is one thing that will be verified. In addition, the process helps prevent human error in cam installation. (We should note that some engine builders opt to wait until final assembly to degree the cam when performing only verification and not alteration; but in our view, it's best to find out whether you have a defective cam or other valvetrain component sooner rather than later, so we highly recommend doing it now!)

There are a few different methods used to check cam timing, but one of the most foolproof is to simply measure the valve events listed at 0.050-inch lift on the cam card and, beyond making sure they are correct, calculate the intake centerline (ICL) from them. It is this process that we go through in the accompanying pre-assembly steps. It's more common to verify using intake lobe specs, but you may also use the exhaust lobe–it doesn't make any difference as long as the specs are listed on your card. Also be sure you make any necessary adjustments for degree wheels that are labeled in 90 or 360 degree increments (ours was labeled in 180 degree increments, with 180 being BDC).

As a final note, all cam degreeing procedures shown in this chapter are illustrated as applied to non-VVT LS engines. Though both the processes of verification and alteration are theoretically applicable to engines equipped with VVT (albeit to varying extents), some modifications to the procedures will be required if you're planning on performing any sort of cam degreeing with such an engine.

Variations on the cam degreeing tools and techniques shown in steps 21-25 are possible. For example, you may wish to substitute a deck bridge with a dial indicator in lieu of a TDC stop. (This item will also help you measure piston-to-deck clearance when calculating compression ratio–see the Appendix). The instructions included with your degree wheel kit will often list other permissible variations, such as performing the process with the cylinder heads already installed.

Keep the following in mind during cam degreeing and other cam-related processes. When both the crank and cam sprocket markings are in the 12:00 position (left photo), this indicates TDC of cylinder #1 at the firing position (between the compression and power strokes). When the crank sprocket marking is at 12:00 but the cam sprocket marking is at 6:00 (right photo), cylinder #1 is at TDC between the exhaust and intake strokes. Some timing sets may be different, and if you have any doubt, call the timing set manufacturer or just watch the movement of the cam lobes to verify.

Performance Tip

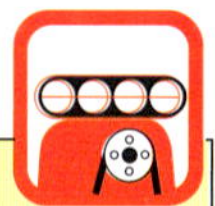

22 Degreeing Camshaft - Find True TDC

We must now determine true TDC for cylinder # 1—this is the industry standard for the crankshaft being at "zero degrees." Install your degree wheel kit's TDC stop. The crank will need to be turned backward slightly in order to drop the piston into the cylinder and allow the stop to be tightened in place (use a couple of old head bolts). Now turn the crank forward (i.e., clockwise when viewed from the front) until the piston hits the stop. Mark the pointer reading on the degree wheel. Then, turn the crank backward until the piston drops to BDC, comes up, and hits the piston stop again. Mark this reading on the degree wheel as well (non-permanent marker preferred!). The halfway point of the two marks you've just made is true TDC, so make any adjustments to the degree wheel's position on the crank to get this halfway point to the zero mark. Verify by spinning the crank forward and then back again, making sure the pointer indicates the same number each time the piston hits the stop (albeit on either side of zero). After being sure your degree wheel is tight on the crank snout, remove your TDC stop.

Performance Tip

23 Degreeing Camshaft - Install Dial Indicator on Lifter

Lubricate and install a lifter into cylinder number one's intake position (nearest lifter bore to the front). Although having the lifter guide tray in place would prevent the lifter from rotating, this is not possible during the cam degreeing process: the tray holds the lifter so tightly, it won't allow it to fall back onto the base circle of the cam once raised. Fortunately, thanks to the roller design of the LS lifter, it actually will self-center itself as you slowly turn the engine—but you should still watch the lifter carefully to ensure it does not rotate out of place and give a false lift reading. Now, mount your dial indicator fixture securely; using a head bolt hole as a mounting point provides excellent stability. Adjust so that the dial indicator's stem is parallel to lifter travel. Also ensure that the dial indicator's tip touches the center of the lifter plunger—an extension will likely be needed, as shown. Zero the dial indicator with the lifter on the cam lobe base circle (this will be the case if you are anywhere near TDC for the firing position).

Performance Tip, Documentation Required

24 Degreeing Camshaft - Turn Crank and Verify Cam Timing

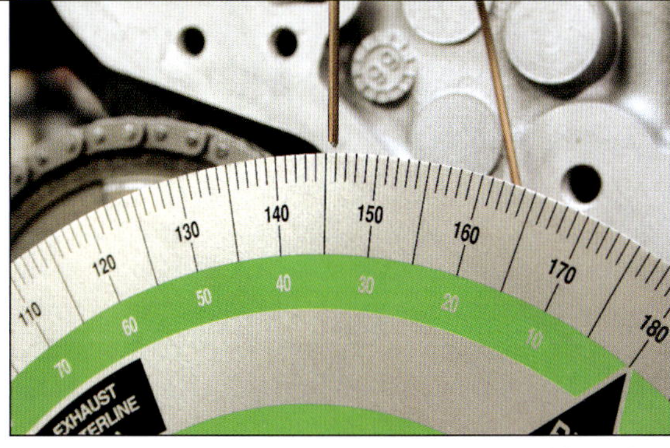

Slowly turn the crankshaft clockwise until the dial indicator indicates 0.050-inch upward movement. Read the number indicated on the degree wheel; it should correspond with the intake valve opening value listed on the cam card (on our cam, -8 degrees BTDC, which is the equivalent of 8 degrees past TDC). Write the degree wheel reading down if it varies from the cam card. Turning the crank again, continue until the dial indicator cycles all the way to full upward lifter movement and then down to 0.050 again. Note the degree wheel reading at this point. In our example, this reading is 146 degrees (shown), which equals 34 degrees past BDC ("34 degrees ABDC"—the same as our cam card). Write this reading down as well if it varies from the cam card.

Performance Tip

25 Degreeing Camshaft - Calculate and Compare

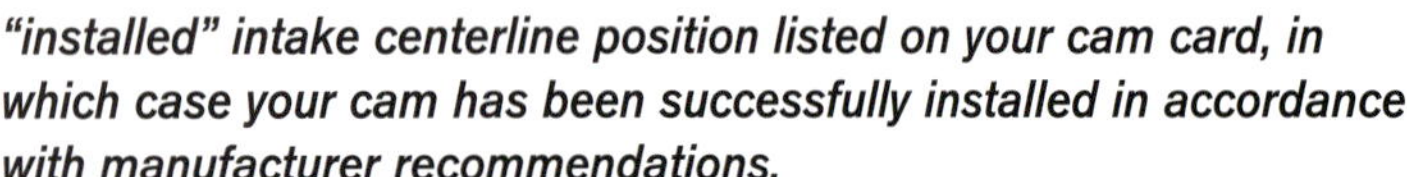

Now do the math—use care in all conversions: add the intake opening and closing degree wheel values you have written down, using absolute values in comparison to TDC. In our example, the two values are 8 and 214 (180 + 34), respectively, and adding them yields 222. Dividing this number in half yields the intake centerline; in our case, 111. The number you come up with should be the same as the "installed" intake centerline position listed on your cam card, in which case your cam has been successfully installed in accordance with manufacturer recommendations.

If there are any variations between the cam card and the above measurements or calculations, repeat steps 21-24 to be sure. If the problem persists, there may be something wrong with your cam or timing set. If you are going to alter cam timing via use of an adjustable timing set, do so now by altering sprocket positioning (shown) and repeating the degreeing process to check your changes. Once you are happy with your results, write down or photograph the respective positions of the adjustable cam or crank sprocket(s) so that you can be sure to install them the exact same way during final assembly.

Special Tool

26 Valvetrain Checks - Measure Valvespring Installed Height

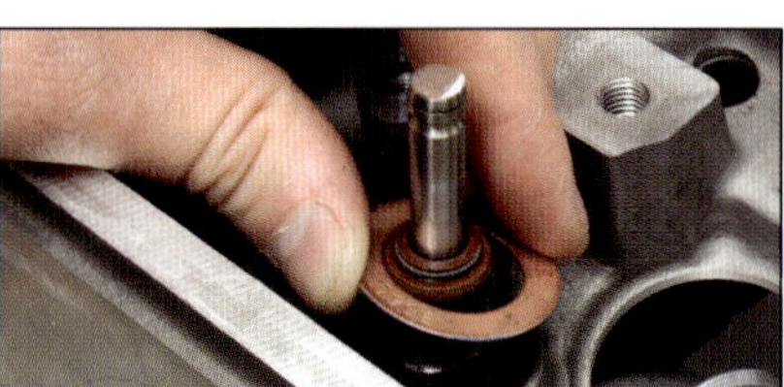

Note: If your machine shop has already assembled your cylinder heads, or if you are using a set of pre-assembled aftermarket heads, skip ahead to step 29. (Your valve stem seals and spring seats must be installed during the next few steps. Refer to step 2 of cylinder head assembly in Chapter 8 for how to do this.) *Insert an intake valve into the cylinder head. Set your valvespring height gauge onto the valvespring seat. After installing a valvespring retainer and locks, twist the height gauge until the valve is held fully closed. Note the reading on the gauge. Valvespring manufacturers normally quote the proper installed height for their springs; stock replacement springs should use GM specifications for installed height. If the height gauge shows less than the required installed height value (or is substantially greater), you will need a different set of springs. If the height gauge reading is greater than this value (but still in the ballpark), you can purchase and add shims under the valvespring to yield the proper installed height. If any shimming is required, you will need to perform these measurements for every valve in each head and keep track of which shim(s) go where so that you can replicate your results during final cylinder head assembly.*

Precision Measurement

27 Valvetrain Checks - Measure Valvespring Retainer to Valvestem Seal Clearance

With a retainer and locks installed on an intake valve, mount a dial indicator fixture on your head. The dial indicator must contact the top of the valve and be parallel to it. Note the indicator reading with the valve fully closed, then open the valve to the maximum intake valve *(not lobe) lift indicated on your cam card (being sure to compensate if not using the quoted rocker ratio). In the case of stock cams, refer to the lift listed in your GM service manual. If the retainer has not yet made contact with the valve stem seal at the point of max lift, push the valve all the way down until this occurs, and note the dial indicator reading. The difference between max lift and the point of contact should be at least 0.050-inch. (Note that during this process, you will need to keep upward pressure on the retainer and locks so that they will not simply fall off of the valve. To make things a little less tricky, high-performance applications may wish to wait until after checker springs have been installed [step 29] to perform this step.)*

Saftey Step

28 Valvetrain Checks - Measure Compressed Valvespring Clearance

You need to double-check that your valvesprings are capable of handling the valve lift that your cam will provide on your application. If your valvespring manufacturer didn't supply a value for maximum spring compression, you can find it easily by placing your valvespring in a vise. Don a good pair of goggles and fully compress the spring until all coils contact each other. Measure the spring–this is the coil bind height. To figure out whether you will approach this value, subtract the intake valve lift on your cam card (again, making any compensation necessary for different rocker ratios) from the valvespring installed height you determined in step 26. This number is your compressed valvespring height, and it should be at least 0.060 inches greater than the coil bind height you've just measured for most springs. Repeat steps 26–28 for at least one exhaust valve.

Performance Tip

29 Valvetrain Checks - Install Checker Springs on Head

Note: This step applies to high-performance applications only. For stock rebuilds, skip to step 30. *If your heads are already assembled, you'll need to remove the valvesprings from the intake and exhaust valves of cylinder number one (follow the appropriate steps enumerated in Chapter 3 during head disassembly). Install checker springs in these locations at this time.*

Important notes:

(1) The checker springs you use must be weak enough so as to not compress the plunger of hydraulic roller lifters. This will mean springs with very little tension at all! As an alternative, you may temporarily install solid lifters in lieu of your hydraulic lifters, but they must be of the exact same plunger height (or at least a known height difference) to properly perform steps 30-34 and the accompanying calculations.

(2) Be sure to also install your lash ("wear") caps at this time if, for example, you are using titanium valves.

30 Valvetrain Checks - Install Cylinder Head on Block

Note: Before performing steps 30-34, see "Determining Optimum Pushrod Length" on page 105. *Make sure you have your lifters for cylinder #1 installed in the block–to avoid having to continually push the lifters back down onto the cam lobe during piston-to-valve clearance checking, high-performance rebuilds should not install the lifter guide tray. Turn the crank as needed so that both lifters are on the base circle of the cam (again, near TDC at the firing position works best). Install head locating dowel pins into the block deck surface and set a head gasket in place on the block. It is best if you have a reusable head gasket for this step (i.e., not the earlier-style graphite-layered GM head gaskets–they will stick to the deck surfaces and be destroyed, making reuse during final assembly impossible). Install your driver side cylinder head. If you will be using stock-style head bolts during final assembly, use an old set of head bolts now. Again, thanks to their torque-to-yield design, new GM bolts can only be fully tightened once. Follow the head bolt tightening specifications and sequence listed in steps 43-44 in Chapter 8.*

31 Valvetrain Checks - Measure Pushrod Length

Insert pushrod length checkers for the intake and exhaust valves of cylinder #1. On most LS engines the rocker pivot supports are not a machined part of the head; in these cases, lay the rocker stand in place on the head—it helps to have a couple of rocker bolts in toward the other end of the head to prevent the stand from shifting. Set the rockers loosely atop the stand and adjust the pushrod length checkers so that they are slightly too short to contact the pushrod seats on the rockers (when the rocker tips are touching the valves). Now bolt the rockers down and adjust the pushrod length checkers so that they just barely contact their pushrod seats (this will be difficult thanks to the minimal exposed area of pushrod provided by LS heads). Jiggle the checkers to ensure they are squarely positioned in both the pushrod seats and lifter plungers below. Since the lifter plungers are not visible in LS engines that have heads installed, it will be difficult to tell whether any hydraulic lifter preload is occurring (there should be none)—this is why so much care is needed when adjusting the pushrod length checkers. Also, if you're using checker valvesprings, use extreme care not to compress them. Once you've achieved pushrod lengths just sufficient to remove all play from the system (and no more), remove and measure the pushrod length checkers. The proper pushrod length for your engine is this measurement plus the recommended hydraulic lifter preload provided by the lifter manufacturer (usually in the neighborhood of 0.060 inch for stock GM lifters). Pushrods are normally sold in length increments of 0.050 inch, so select the length that puts the lifter preload within its acceptable range. Please note that results may differ slightly between the intake and exhaust valves, and if the difference is significant, you may consider using a different pushrod length for each. Repeat the measurement process to ensure accuracy—it doesn't hurt to do it on at least one other cylinder, too (especially on the other cylinder head if head resurfacing has been performed)!

Performance Tip

32 Valvetrain Checks - Install Pushrods and Rocker Arms

Note: steps 32-34 apply to high-performance applications only. For stock rebuilds, skip to step 35.

Because checker springs are being used in lieu of actual valvesprings, use of your new correct-length pushrods is not possible during the next few steps. The reason for this is that correct-tension checker springs do not provide enough force to compress the hydraulic lifter plunger, so your new pushrods would actually cause the valves to hang open slightly. For this reason, use your pushrod length checkers set to the correct length minus the hydraulic lifter preload. (If you find that your checker springs are strong enough to compress the lifter plunger, consider installing a set of solid lifters, which you would leave in from now through step 34.) Before installing anything onto the head, though, spin the crankshaft over several times to allow the lifters to re-align themselves on the cam lobes (remember, you should not have lifter guide trays installed, and the lifters may have rotated during pushrod length measurement. This is OK with the lifters on the base circle of the lobes, but it will not be OK now!). Then install the pushrod length checkers and rocker arms for cylinder #1 (ensure the rocker stand does not twist on the head while doing this). If you are using aftermarket adjustable rocker arms, you will need to set the lash adjuster to a position that removes all lash but keeps the valves closed.

Determining Optimum Pushrod Length

The process of determining the correct pushrod length to use in your engine is critical in the assembly of any LS, stock rebuilds included. Here's why: as we mentioned in Chapter 1, the Gen III/IV uses a valvetrain with a "net build lash scenario," which basically means that it's non-adjustable. When GM rockers bolt to the cylinder heads, proper lifter preload is provided by the pushrods being slightly longer than the distance between the uncompressed lifter plunger and the pushrod seat in the rocker–and there is no way to fine-tune it.

Think about all of the variables here: even when reusing stock parts, cylinder head and block resurfacing or use of a different thickness head gasket will cause pushrod length needs to change. Swapping GM cams between engines can also be a factor because they can differ in base circle diameter. And aftermarket parts like different lifters, rocker arms, or cylinder heads (with or without different valve angles and lengths) really throw a wrench into things.

Since piston-to-valve clearance is being established on high-performance engines during the pre-assembly process, it only makes sense to figure pushrod length at the same time. The good news for stock or near-stock applications that do not need to check piston-to-valve clearance is that you can perform pushrod length measurement later during final assembly. Pushrods of nearly all lengths are readily available and getting a hold of them should only hold up your rebuild a day or two. This avoids wasted time in torquing bolts, having to install valvesprings on as-yet-unassembled heads, and possibly damaging non-reusable head gaskets now. If you choose this route, you may skip steps 30-34 entirely. (As an alternative, the measurement can also be taken now without the head gasket in place and adding its known compressed thickness to pushrod length–don't torque the head bolts, and be sure to drop the piston into the cylinder a bit, if you do it this way!)

We're going to show you how to determine pushrod length using stock-style non-adjustable rocker arms. For aftermarket adjustable rockers, instructions on pushrod length measurement should be included with the product instructions. Procedures are similar whether using hydraulic or solid roller camshafts, though hydraulic roller lifters are used in our example.

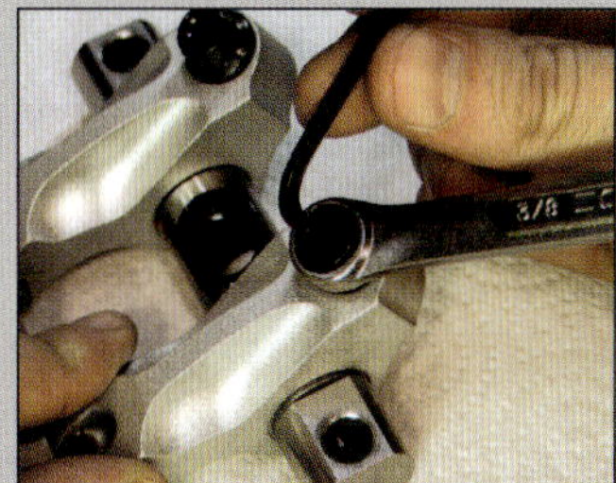

Even when using an aftermarket adjustable rocker system, determining the correct pushrod length is still very important for proper valvetrain geometry. For most styles of adjustable rocker, the process is the same as we show you here, except you will simply need to lock the lash adjuster near the center of its travel (see the instructions included with your rockers for complete details). We will show you how to install and adjust shaft rockers like these Jesel units during final assembly in Chapter 8.

33 Valvetrain Checks - Inspect Rocker Geometry

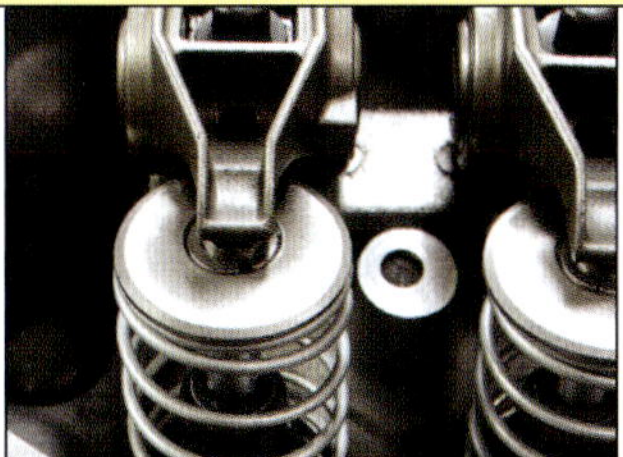

Slowly turn the crankshaft over and watch the movement of the rocker arms. (If any resistance to movement is felt, stop: the valves may be contacting the pistons. If this is the case, you definitely do not have adequate piston-to-valve clearance. See step 34 how to address this problem.) A main area to check is interference between the underside of the rocker and the valvespring retainer (pen pointing). Fitment problems here are more likely if using valvesprings with wider-than-stock retainers (as shown), i.e., those for use with non-tapered valvesprings. One other noteworthy area of rocker geometry to look at is where the tip sits on the valve stem. It should be nearly centered when viewed from above, and beyond this, look at the rocker from the side as it traverses through low, mid, and high lift. The tip should be contacting the center of the valve stem during the mid-lift period. If using aftermarket rockers, some additional checks can be performed at this time (some of which will need to be done on more than one cylinder); for example, you can inspect for interference between the rockers and valve cover rail or valve cover. Minor massaging of the valve cover rail may be needed with some rocker systems (shown). You can also set your valve cover atop the head (sans bolts and gasket) and turn the engine over. If the valve cover moves or you hear any "clicks," the rockers are contacting the underside of the valve cover and modification may be required.

Performance Tip

34 Valvetrain Checks - Check Piston-to-Valve Clearance

Rotate the crankshaft so that cylinder number one's intake valve is closed. Mount a dial indicator so that it touches the top of the intake valvespring retainer and is parallel to the valve stem. Rotate the crankshaft until you see the intake valve just begin to open; note the indicator reading at this point. Using your hand, rotate the intake rocker until the valve contacts the piston (the valve may have to be opened very far for this, or it may never contact the piston at all, particularly toward the beginning of the opening event). Write down the dial indicator reading and release the rocker. The difference between the two indicator readings is your piston-to-valve clearance at that crank position. Now rotate the crank a few degrees and repeat, again writing down the dial indicator reading if and when the valve contacts the piston. Do this until the intake valve has completely opened and closed, then set up the dial indicator on cylinder number 1's exhaust valve and repeat the entire process. In most cases, minimum piston to valve clearance should be 0.100 inch for the intake and 0.125 inch for the exhaust valve (bare minimums should be 0.080 and 0.100 inch, respectively). If clearance is tighter than this, you may be able to alter cam phasing slightly to accommodate, or you can swap to a different cam. If this is not an option, you may have to machine the tops of your pistons to provide deeper valve reliefs.

Professional Mechanic Tip

35 Piston Ring Fitting - Setting Compression Ring End Gaps

Remove the cylinder heads, camshaft, all parts of the rotating assembly, bearings, and all other parts you may have installed until you are left with only the bare block. Starting with cylinder number 1, slip a top compression ring in and use your ring squaring tool to get the ring completely horizontal in the cylinder. It helps to pull up on the ring from underneath to get it flush with the bottom surface of the tool. Then, remove the squaring tool and measure the ring end gap with a feeler gauge. If you are using a drop-in ring set, the gap should match that given by the manufacturer or listed in your GM service manual. If using a file-fit set, the gap should be significantly tighter than you need it to be at the moment. To widen the gap, use your ring filer. Grind against only one side of the ring, being sure to hold it squarely against the filing wheel. Parallel ring ends are crucial to the longevity of your engine! The coarse grit of the filing wheel will remove material quickly, so take care not to go too far. Reinstall the ring in the cylinder and measure the end gap again; repeat the filing-squaring-measuring process until you've got the gap right. Then move to the next cylinder and continue until you have fit (or verified) top ring end gaps for them all. Repeat the entire process for the 2nd rings, performing fitting for each cylinder sequentially. (Hint: leaving each ring you've just done in its bore prevents you from accidentally mixing the rings up or cutting them for cylinders you have already fit rings to.)

The Science of Ring Fitting

We've gone through a number of procedures during the pre-assembly process, but we are leaving ring fitting until last for one simple reason: it gets tiny metal shavings everywhere! (Actually, this is not true if you're using drop-in rings, but you must still verify the end gap of the top and 2nd rings for each cylinder if using such a pre-fit set.) Remember, it is the end gap of the top and 2nd compression rings that we're interested in when ring fitting; oil rings (oil control rails/ expanders/ supports) do not require end gap adjustment.

Ring fitting custom-tailors the end gap of each individual compression ring, and it's necessary not only because cylinder bores can vary slightly, but because piston ring end gap needs are unique to a given engine. Indeed, properly setting piston ring end gap is a crucial part of any engine build, and it all starts with determining exactly what this end gap should be. Different ring designs and materials respond differently to conditions inside the cylinder, and piston design also factors into how much heat the rings actually see. To further complicate matters, the actual ring gap is not even constant while the engine is running. For example, when the engine is started cold, ring end gap is significantly wider than it would be during full-throttle operation.

Simply put, a compression ring requires a sufficient gap to allow the ring room to expand with heat slightly during engine operation, thereby preventing the ends of the ring from butting together (a scenario that would result in severe parts destruction inside the cylinder). On the flip side, too much gap will cause engine efficiency to be lost, as an excess of gases will escape from above the piston into the crankcase—gases that would otherwise have added to cylinder pressure. It is for these reasons that it's an exceptionally bad idea to deviate from what a ring manufacturer recommends regarding end gap.

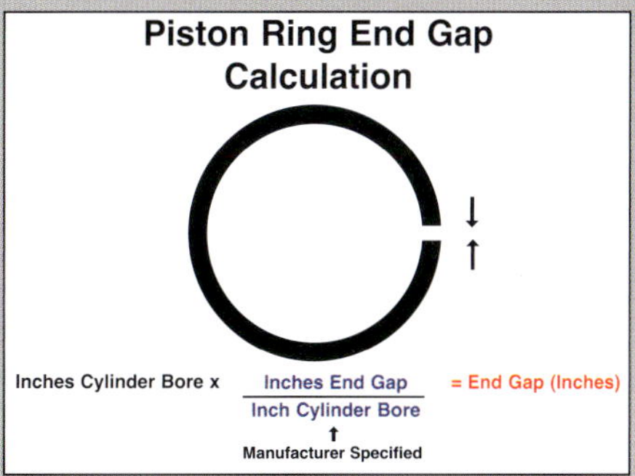

Let's take an example. A typical set of high-performance rings consists of plasma-moly top rings and ductile iron 2nd rings, and many manufacturers of such a ring set will specify a minimum of 0.004 inches of end gap per inch of cylinder bore diameter for the top ring. Let's say we're putting together an engine with a 4.005-inch bore. The calculation for end gap would be as follows:

4.005 inches bore x 0.004 = 0.0160 inches ring end gap

It should be noted that this number would be the minimum end gap for a typical naturally-aspirated engine. Other engines require somewhat wider end gaps, as the increased cylinder pressures that come along with nitrous oxide, turbochargers, or a supercharger add heat inside the cylinder, causing the rings to expand even more. Further, gap recommendations are usually slightly larger for 2nd rings than for top rings, as many engine builders feel a wider 2nd ring gap prevents pressure buildup between the rings (though some actually recommend that its gap be narrower since this ring experiences less heat). Again, ring design and material play a role in all of this.

We can't stress this enough: always stick to the end gap recommendations given by the ring manufacturer. If you have any questions that aren't explained in the instructions (or if your rings didn't come with instructions), call the manufacturer for clarification. Remember: botch the ring fitting step, and the result will either be loss of engine efficiency or loss of an engine!

36 Piston Ring Fitting - Deburr and Label Compression Rings

After ring fitting is complete, it's important to deburr the ends of each ring that has been file fit using a stone or similar instrument. This removes any stray pieces of metal that could score the cylinder or become lodged between the piston and the ring, thereby preventing the ring from rotating or expanding properly against the cylinder wall. (Note that it is not a bad idea to inspect the end gaps of drop-in rings to determine whether the deburring process might be needed on them, too.) Deburr one cylinder at a time so as to avoid any chance of ring mixup—each is now uniquely fit to a particular cylinder! When both the top and 2nd rings have been deburred, clean the rings and label each cylinder set (top and 2nd ring) with a magic marker and masking tape.

Pre-assembly Complete

The most tedious part of any engine build is now over! At this point, you should discuss any unresolved issues you may have encountered during pre-assembly with your machine shop. With any such problems having been taken care of, remember that you will need to clean all engine components (as detailed in Chapter 6) one more time before proceeding to Chapter 8, Final Assembly.

FINAL ASSEMBLY

It's been a long road to get to where you're at right now. Think about it: you came up with a plan of action for your rebuild before even turning a single wrench. You acquired the tools that were needed for the project, then set about the dirty process of engine disassembly. You then inspected your engine parts, selected others to purchase, and had machine work performed on many of them. You cleaned all components and then verified proper fitment of all parts during pre-assembly (not to mention took care of any issues that may have arisen during that process).

We'll, it's *finally* time for the main event: engine assembly. The good news is that we're on easy street from here on out, and though it may sound ironic, you'll probably find the final engine assembly process downright simple in comparison to some of the tasks you've performed thus far. Not that this is a time to completely relax; a great amount of care and attention to detail is still required! Nonetheless, chances are that this is the phase of your rebuild project you will enjoy the most, as all the hard work you've done thus far will, at long last, begin to pay off!

Before You Begin

At this point, a few last words are appropriate. Before beginning final assembly, you should have all of the applicable tools listed in Chapter 2 in hand and ready to use—read through this chapter first to see which ones you will need if you're unsure. Also, it should go without saying that before getting started, you should have performed *all* applicable pre-assembly procedures listed in Chapter 7 and had any and all additional machine work performed on your parts, as appropriate. And even though you went through the component cleaning process (Chapter 6) before embarking on pre-assembly, you will need to have cleaned all of your engine components *one last time* before embarking on final assembly. Remember, contamination left on parts can ruin your engine!

Final Assembly Supplies

The following supplies are either necessary or helpful during the final assembly process, and they apply to several steps. Without further ado, they are:

Assembly lubricant

All friction surfaces inside the engine will, of course, require lubrication so that they are adequately protected at startup. Though a thick engine oil is acceptable to use on bearing surfaces, additional protection can be had with a specialized assembly lubricant designed with bearings in mind. Another type of assembly lube is so-called high-pressure lube, which is designed to be applied at points where significant friction occurs, such as the pushrod cups of rocker arms (this stuff is sometimes moly-based and is usually included with aftermarket rocker arms). Use of all the right lubrication in all the right places will ensure a trouble-free break-in and long life for your LS. Also note that the term "assembly lube" can also apply to the lubricants used on the threads of some aftermarket fasteners, and as mentioned previously, small packages of this stuff are normally included by the fastener manufacturer. Refer to the product information for any assembly lubricants you pick up so that you're familiar with the approved uses of each before applying.

A dedicated bearing assembly lube (top left) will help protect your bearings and crank journals during initial startup, while a high-pressure lube (bottom right) will do the same for your pushrod seats. The Royal Purple assembly lube at top right "bridges the gap" between bearing lube and high-pressure lube, and can be used on both types of surfaces. Finally, fastener assembly lube (bottom left) assists in achieving proper torque readings, and this stuff is nearly always moly-based.

Engine oil

Regardless of whether you're using purpose-formulated assembly lube on your bearing surfaces, engine oil will still be needed during final assembly. Usages range from lubrication of certain fasteners (for proper tightening) to coating of surfaces like cylinder walls, lifters, and cam lobes. Thicker-weight conventional oils such as SAE 30 are preferred. Do *not* use thin weight oils such as 5W-30: a thick enough layer of it may not cling to metal surfaces between now and startup! Also, despite what you may have heard about GM putting together some of its high-output LS engines using *synthetic* oils, they are not recommended for our purposes.

An oiling can will prove useful when applying engine oil to your parts so that they (and your hands) stay as clean as possible. A small set of brushes is another option, as we talked about in Chapter 7.

Threadlocker, thread sealant & anti-seize

Certain fasteners and engine block plugs will require thread locking or Teflon-based sealing compounds. Each of these will prevent loosening over time and unwanted fluid seepage, respectively. Note that while you should pick up some anti-seize compound, this stuff should only be used on spark plug threads.

RTV Silicone

Assembly of a Gen I or II small-block would require an entire tube of this stuff or more. Fortunately, modern gasket technology means that only a few dabs are needed when bolting together an LS, so a small tube is all you'll need.

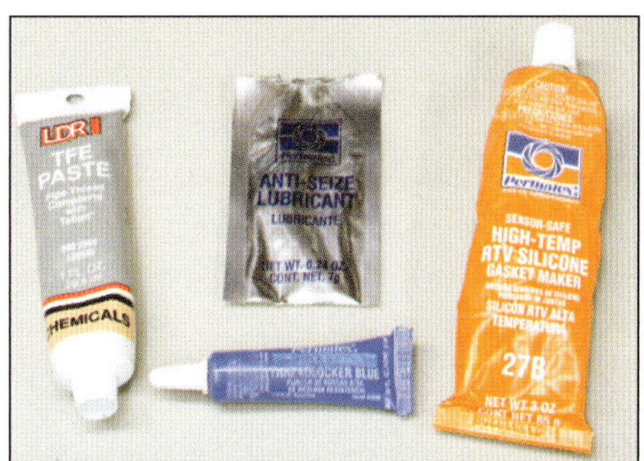

You can't put together an LS without these: thread sealant (also known as pipe thread compound or "Teflon paste"), threadlocker (medium-strength recommended), RTV silicone, and anti-seize.

Parts organizer trays

Though not an absolute necessity, it's nice to have something to safely hold your soon-to-be-installed engine parts while you are going through the assembly process. Parts organizer trays are great because not only do they make items less likely to fall off of your worktables, but they keep numbered parts in order, helping prevent inadvertent mixups of cylinder-specific items like pistons and rods.

Goodson is a great source for parts organizer trays like this so-called block organizer. Piston/connecting rod assemblies are especially likely to tip over once assembled, so why not have them lying safely atop your worktable?

Supplies to stay clean

We've already reiterated that your parts must have been cleaned following the procedures in Chapter 6 before beginning final assembly (and frankly, you're probably sick of hearing it). But you'll also need to make sure they stay that way as your engine comes together. Hand cleaner should be kept nearby to clean your paws frequently during assembly. Some of the usual cleaning supplies we've seen before will be needed as well, particularly assembly wipes (which should always be used unless otherwise specified), which will allow you to give the surfaces of engine parts one last wipe before they're hidden inside your LS forever.

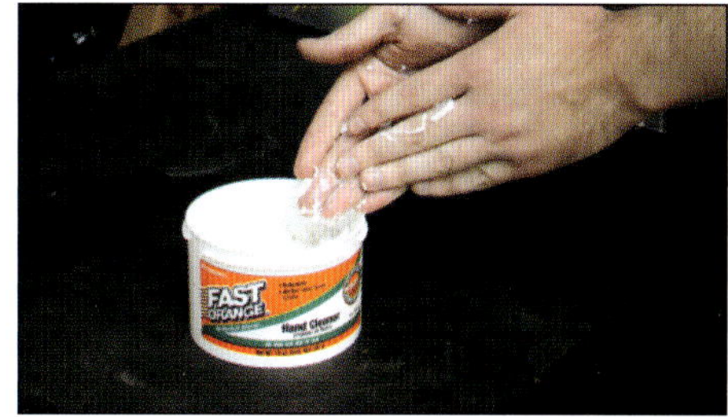

While you've hopefully been keeping your hands clean throughout all stages of your rebuild project, it's more important than ever to do so now. The life of your LS depends on it! You may also consider wearing disposable rubber gloves during the final assembly process.

Final Notes

Before getting started, one last note: it should be obvious that there will be a lot of bolt tightening to perform during final assembly. You will need to have all fastener tightening specs—torque, torque angle, bolt stretch, etc.—ready and verified before beginning. The reason for this is that although we're quoting typical GM specs in this chapter, you must always use the specifications supplied by the manufacturer when using aftermarket fasteners (as well as any special instructions supplied with any aftermarket parts). *Do not apply lubricants, sealants, or threadlocker to any fasteners unless specified herein or by the fastener manufacturer!* In addition, the final caveat is that since there are so many variants of the LS engine architecture (a pool which continues to increase by the model year), it would be impossible to guarantee that all fastener tightening specs quoted herein are correct for each and every LS engine. We will attempt to note any such differences as we go along, but it doesn't hurt to have your GM service manual at hand to verify or look up values unique to your engine.

Step-By-Step Final Engine Assembly Procedures

1 Ready to Begin Assembly

The most exciting time of the engine build is now upon you, as all of your shiny new or refurbished parts will come together to make what you've been dreaming of: a fresh LS engine! Don't let the adrenaline get the better of you, though: follow the below steps carefully (and in order) to ensure you don't make any mistakes.

2 Install Screw-In Engine Block Plugs

Flip the engine block upside down on its stand. Install any and all block plugs of the screw-in type, including oil gallery and coolant plugs. You should have noted where they all came from during disassembly. All of these are safe to reuse if their sealing washers are in good condition, but be sure to use Teflon-based thread sealant on the threads and under the sealing washer (new plugs include sealant on them). Torque this style of plug to 44 ft-lbs, or 30 ft-lbs if one of these plugs is an engine block heater.

3 Install Press-In Engine Block Plugs

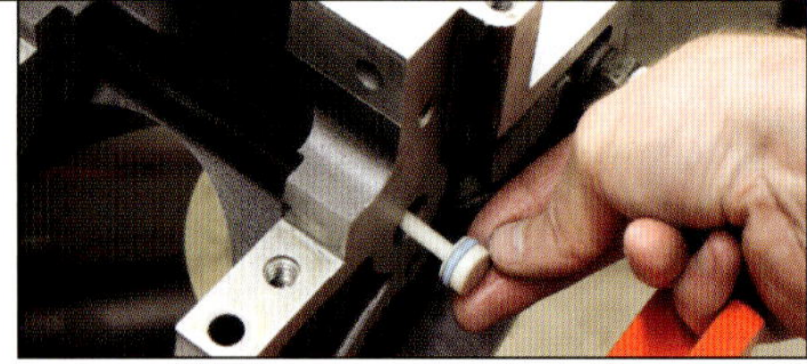

Install a new front oil gallery plug into the front driver side of the engine block. Apply medium-strength threadlocker to its circumference, and tap it in place using a flathead punch or similar instrument. Its outer lip should be recessed about 0.01 inch when fully installed. At this time, also install the rear oil gallery plug (a.k.a. barbell restrictor; use of a new one is strongly recommended). This item can simply push into place by hand, and will protrude just slightly when installed properly. Use no sealant on this plug, just make sure its O-ring is intact and lightly lubricated with engine oil.

4 Install Main Bearings

Wipe each bearing shell seating surface in the block or main cap, as well as the back side of each shell, before proceeding. The upper main bearing shells (grooved) must now be installed into the engine block and the lower main bearing shells (solid) into the main caps. Be sure to install the thrust bearing shells into the center block bulkhead (bottom of first photo) and #3 cap. If you established the need to make one or more of these bearing shells undersize or oversize during pre-assembly, ensure each is being installed in its correct location. Each shell has a tab that fits into a recess in the cap or block, preventing backward installation (finger pointing in second photo). Make sure each shell is fully seated with its edges flush with those of the block or main cap–crooked bearings could spell disaster! Also ensure that your main cap locating dowels are in place in the block at this time (only certain engines have these; for example, the LS7).

Professional Mechanic Tip

5 Insert Crankshaft Into Block

Give the seating surfaces of the block and main caps, as well as the bearings themselves, a final wipe. Lubricate the upper main bearing shells. Do not forget to lubricate the thrust faces of the #3 main bearing! Spread some lube on the crank's main journals, too. Now, grasping the crankshaft by the snout and rear flange, slowly lower the crank into the block. Use care not to nick any of the crank journals. Just as the crank is about to rest on the bearings, it may get stuck; slight adjustments in angle of the crank will likely be needed to get it to seat, as it will only go in just the right way. If gentle wiggling does not coax the crank to seat, pull the crank slowly upward and start over; excessive jostling will damage the crank journals or bearings. It may help to rotate the crank slightly as you lay it in.

6 Install Main Bearing Caps and Bolts

After lubricating the lower main bearing shells, lay the main bearing caps in place, being sure to install each numbered cap in the correct location. The "wings" at the edges of the caps all face toward the rear of the engine, with the exception of the #5 cap (you can see this in the accompanying cap numbering photo). The caps likely will not seat fully by hand, as remember they are a tight fit between the deep-skirt oil pan rails. Insert the M10 main cap bolts and start them by hand. To guide the caps down, alternate tightening the bolts side-to-side to ease each cap all the way into place (don't do this on caps that use locating dowels—tap the cap in place with a rubber mallet). The longer, non-studded main bolts go toward the center of each cap, while the studded ones go toward the outside. Only snug these bolts for now, then install the side cap bolts loosely. If you are re-using your old side bolts, apply some RTV sealant under the heads to prevent oil leakage.

Important Notes:

(1) Though GM recommends these M10 bolts be installed dry, some engine builders use a small amount of oil on the threads to help prevent any possibility of thread damage during tightening (you can put some under the heads of the bolts, too). This is only permissible because these bolts use the torque-plus-angle method; doing this would destroy a proper reading if relying on a torque spec only!

(2) If not using GM bolts, follow the lubrication instructions provided by the manufacturer.

Torque Fasteners 

7 Tighten Inner Main Bearing Cap Bolts

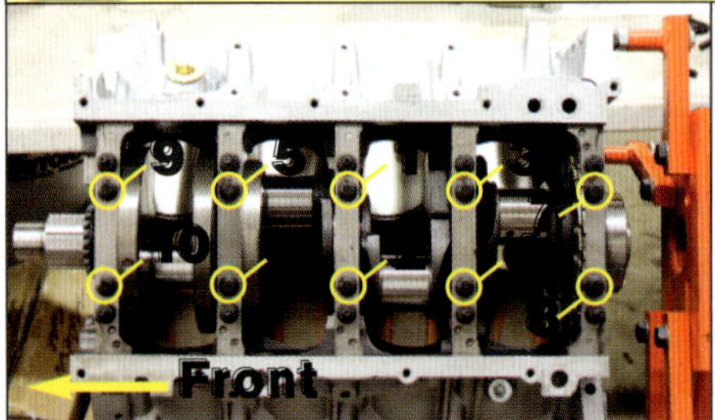

Torque all M10 inner main bolts (numbered 1-10 in the accompanying diagram), in the sequence shown, to 15 ft-lbs. Before proceeding further, you must use a rubber mallet to hit the crank rearward, then forward, with a rubber mallet. This aligns the thrust bearing surfaces and it is important to note that final thrust of the crank must be in the forward direction! Then use your torque angle gauge to twist these inner main bolts in sequence an additional 80 degrees. (See "Utilizing Fastener Stretch" on page 118 for reasons why this "torque plus angle" methodology is used by GM.)

Torque Fasteners 

8 Tighten Outer Main Bearing Cap Bolts and Side Bolts

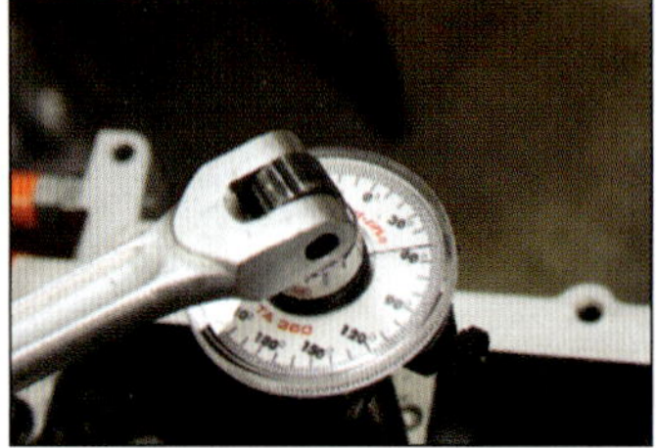

Now, torque all M10 outer main bolts (numbered 11-20 in the first picture), in sequence, to 15 ft-lbs. Once this is done, use your torque angle gauge to add an additional 51 degrees of twist to each outer main bolt, again in sequence. With all M10 main bearing bolts now tight, the main cap side bolts can be addressed. Torque each to 18 ft-lbs; there is no required sequence for these bolts save to say that you should tighten one side bolt and then the other before moving to the next cap. There is also no angle to add to the side bolts.

9 Assemble Pistons to Connecting Rods

Unless you are using a press-fit piston pin (in which case your machine shop will already have put together your piston/rod assemblies for you), now is the time to assemble your pistons to your connecting rods. Read "Piston and Connecting Rod Orientation Rules" on page 113 before proceeding. Piston pin retaining clips vary in style; some are c-clips that install with snap ring pliers or similar tools, while others are spiral-type and must be stretched open to ease installation (shown in hand, such a spiral-type lock must then be worked into one side of the piston pin bore by simultaneously rotating and pushing it into its groove). Once you have a clip in one side, lightly lubricate the piston pin as well as the friction surfaces in the piston and rod with clean engine oil. Slide the pin through the bores in the piston and rod until it hits the clip on the far side of the piston. Then install the other clip atop the pin. Make sure these pin retaining clips are fully seated—they'll normally click when they're all the way in (a flathead screwdriver comes in handy for this)! Repeat for all pistons and rods, remembering to install any piston "notched" for reluctor ring clearance onto your #8 connecting rod (see Chapter 5)! At this point, we suggest marking the face of each piston with magic marker to correspond with any number previously scribed onto the rod.

Piston and Connecting Rod Orientation Rules

Determining proper piston and connecting rod orientation on the Gen III and IV is critical because of factors like the offset pin design of factory (and many aftermarket) pistons and the filleted journals of many aftermarket cranks. To ensure that you assemble your pistons to your rods–and, subsequently, install them into the engine–correctly, take note of the below orientation rules. In the case of pressed piston pins, these rules will also help you verify that your machine shop assembled your pistons to your rods correctly.

Pistons

Inspect the face of your pistons for an imprinted arrow. Without exception, this arrow must point toward the front of the engine. Similarly, some pistons are manufactured with a small mark toward one side of the piston. This mark denotes the side of the piston that must face the front of the engine. (Sometimes there is a small chunk of material on the underside of the piston, on or near the pin bore, that serves the same purpose; but rarely will this be your only indicator). Additionally, if your pistons have two side-by-side valve reliefs, note that they must face the top of the engine, i.e., be installed toward the intake manifold. Those with pistons having four valve reliefs (or one valve relief on either side of the piston pin centerline) don't have to worry about this last point.

Connecting Rods

Gen III engines

These rods have one side that is flattened in the vicinity of the bolt flange. This flattened area must always face the front of the engine (and so will also coincide with any mark/arrow on the piston). For LQ9 engines–the only Gen III that featured a floating piston pin–follow the rules for Gen IV engines below.

Gen IV engines

Look at the area where the rod cap meets the rest of the rod. Near these mating surfaces, you will see a wider "tab" on one side of the rod than the other. It is this wider tab that faces the front of the engine. (Some Gen IV rods use a tab on one side and none on the other; in this case, the tabbed side faces forward.)

Aftermarket rods

When using aftermarket connecting rods, as with aftermarket pistons, always follow the orientation instructions supplied by the manufacturer. As a general rule, aftermarket connecting rods have a bearing bore shape designed to match up with the large-radius fillets of most aftermarket crankshafts.

Arrows and markings like these indicate which side of the piston must face the front of the engine. Pistons without such markings can be installed any way, with the exception for valve reliefs noted. In the case of the upper photo, these pistons have both an "F" marking and two side-by-side valve reliefs each. The piston toward the top of the photo would be installed in a driver side cylinder (1, 3, 5, or 7), while the piston toward the bottom of the photo would be installed into a passenger side cylinder (2, 4, 6, or 8).

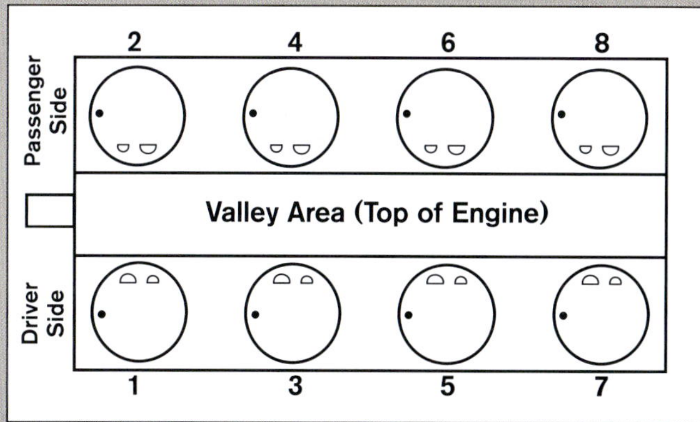

This diagram shows piston mark and valve relief orientation as well as the cylinder numbering system GM uses (and which we use throughout the book as well). When correctly installed, the pistons' marks or arrows will all face the front of the engine. Also, the intake valve reliefs (which are larger than the exhaust valve reliefs) will face the front of the engine on odd-numbered cylinders and the rear of the engine on even-numbered cylinders. Please note that though some pistons are manufactured in such a way that they may have additional, "vestigial" valve reliefs facing away from the intake manifold (on the far side of the piston pin centerline), you should only concern yourself with valve reliefs that are facing the intake manifold.

Piston and Connecting Rod

CONTINUED

Gen III rod orientation. The flattened areas on all rods must face the front of the engine once installed. You can see this on the rod on the left.

Gen IV rod orientation (typical). The side of the rod with the wider tab must face the front of the engine once installed. You can see this on the rod on the right.

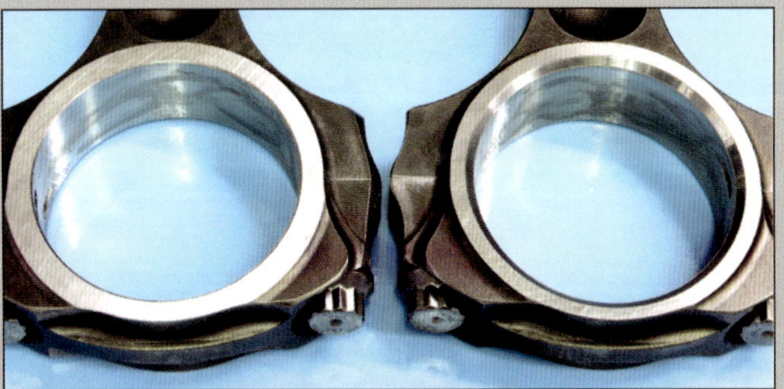

A close inspection of a typical aftermarket LS rod reveals that the two sides of the bearing bore are not the same; one is more heavily chamfered (right), and it is this side of the rod that must normally face the crank journal fillet (i.e., toward the intake valve side of that cylinder). Please note that many GM rods appear to have this same difference in bearing bore chamfers side-to-side; ignore this in the case of factory rods!

10 Determine Piston Ring Clocking

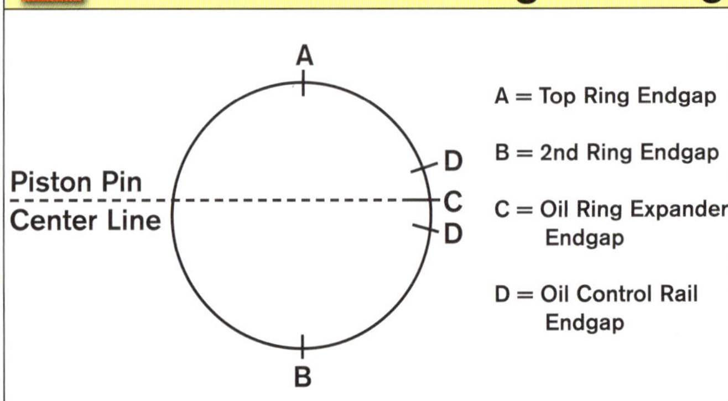

Most piston or piston ring manufacturers will provide specific guidelines as to ring "clocking," or placement of the end gaps about the piston's perimeter. If no clocking recommendations have been specified, you should follow the ring clocking diagram shown here during the below steps of piston ring installation (compression rings can easily be repositioned once installed on the piston, while oil rings are a little more tricky to slide against one another). The most important thing to note is that the oil control rails must be placed about 1-inch to either side of the expander end gap. Most modern ring packages don't require precise clocking other than this, but you should still space the end gaps of the top and 2nd rings approximately 180 degrees apart.

Performance Tip

11 Install Oil Ring Support

Some high-performance LS pistons with a short compression height have an oil ring groove that intersects the piston pin bore (see the Appendix). Such pistons require installation of an oil ring support, which sits beneath the oil rings and provides the necessary structure in the area of the piston pin. An oil ring support must be installed before any piston rings, and this is most easily done using ring expander pliers. Oil ring supports will often have a dimple (pointing) that must face down in the area of the piston pin bore. This dimple prevents the support from rotating out of place while the engine is running—you do not want the gap in the support entering this area!

12 Install Oil Rings

It is easiest to install rings if your piston is standing upright (so that you have two hands to work with); your rod vise lying flat on the surface of a table works well for this purpose. Beginning with the #1 piston, lightly lubricate the surfaces of all rings with clean engine oil. Install your oil ring expander into the bottom ring groove, which is the wavy-looking ring (it may have a piece of thin wire connecting its ends, like this one). Note that some ends of expanders simply butt together, while others lock in place. Then install one oil ring control rail below the expander and another oil ring rail above the expander. Oil ring control rails are easily installed by hand using a light twist (set one end in and hold it with your thumb), and ring expander pliers are not required. Note that if your piston has a small opening beneath the oil ring groove (near left thumb in second photo), you should move the rail endgaps past this area, i.e., further apart than just the 2:30 and 3:30 positions shown in step 10.

Special Tool 

13 Install Compression Rings

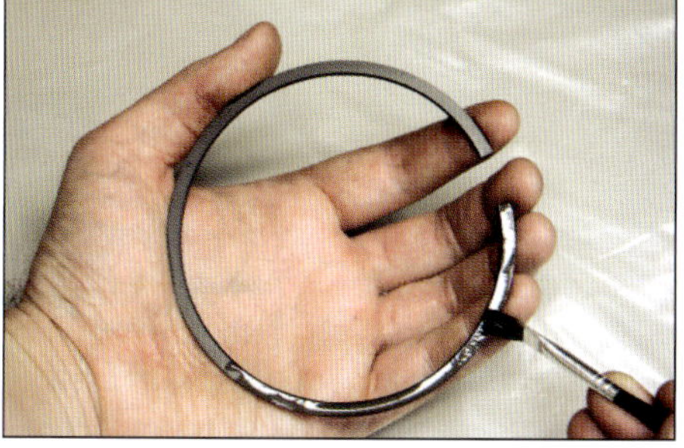

Proceed to lightly lubricate and install your compression rings one at a time, being sure you are using the appropriate compression ring pair for the #1 piston (they should have been marked after ring fitting). Start with the 2nd ring, which sits in the lower of the two compression ring grooves and normally has a duller finish. Be sure to install the correct side facing up: this will be noted on the instructions included with your rings and is normally indicated by a small dot or a beveled/grooved inner or outer edge (which may face either up or down depending on your ring set). Use ring expander pliers to expand the ring just enough so that it slips around the piston; too much can damage the ring. If you do not have ring expander pliers, you can use your thumbs to push the end gap apart. Now install the top ring, again making sure to follow the ring manufacturer's instructions on any bevel or dot placement. Once in their grooves, the compression rings should stick out a bit; this is normal and will help the rings put tension on the cylinder walls. Repeat steps 11-13 (as applicable) for pistons #2 through #8.

Professional Mechanic Tip 

14 Install Crank Turning Tool

PRO TIP As with pre-assembly, a lot of crankshaft turning is required during final assembly. For this reason, it helps to install your crankshaft turning tool (if you have one) onto the crank snout at this point. Depending on the style of your tool, you may have to install your oil pump drive gear (which may be part of your crank sprocket) onto the crank snout before putting the crank turning tool on—see step 27. Again, if you do not have a crank turning tool, we recommend simply installing your old crank bolt and using a 24mm wrench to turn it.

Critical Inspection 

15 Install Connecting Rod Bearings

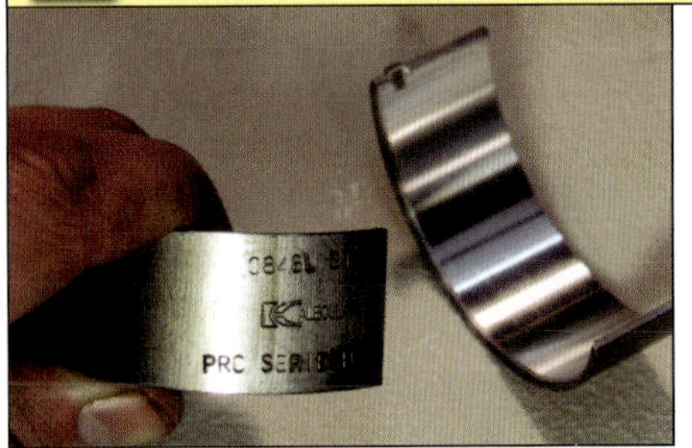

If your rod caps are currently installed tightly to your rods, remove them in accordance with "Proper Connecting Rod Cap Removal" on page 96. Start with the piston/rod assembly for cylinder #1 (remember, each is unique!). After ensuring the bearing shells and their seating surfaces on the rods are clean, install the lower connecting rod bearing shell into the rod cap and the upper shell into the top portion of the rod. As with main bearings, each bearing shell will have a tab that will fit into a recess in the rod or cap, and the shells must be fully seated with their edges flush with those of the rod or cap. Also note that rod bearing shells are not always the same, and your rod bearings may have a top and a bottom half. This is especially true when using chamfered rod bearings to match high-performance filleted cranks. If this is the case, check the back of the bearing shells, which are often stamped with a "U" or "L," indicating upper and lower shells. As a double-check, watch that the chamfer in the edge of such a bearing will face the filleted edge of the crank journal (i.e., the chamfer must face the front of the engine on odd-numbered cylinders and the rear of the engine on even-numbered cylinders). Lubricate the bearing shells with assembly lube.

Professional Mechanic Tip 

16 Turn Engine on Stand and Turn Crank to BDC

Turn the engine on its stand so that the cylinder deck surface is as horizontal as possible. Give the appropriate cylinder wall one final cleaning, and lightly lubricate it with clean engine oil using a lint-free towel or assembly wipe. Then turn the crankshaft so that the rod journal for the cylinder you are working on is at its furthest point below that bore (bottom dead center). This will give the most room to guide the rod onto it. Also lubricate the crank rod journal with assembly lube.

Special Tool 

17 Adjust and Prepare Ring Compressor

No matter what type of ring compressor you are using, lightly lubricate its inside surface with engine oil. If you have a tapered-sleeve ring compressor, set it roughly atop the bore. Adjustable sleeve-style ring compressors need to be adjusted until the cylinder liner can no longer be seen around its inner circumference. (Many machine shops make a slight chamfer in the top of the bores to ease ring installation, and you can see this in the first photo). As for band-style adjustable ring compressors, they should be wrapped snugly (not tightly) around the piston at this time, leaving the piston skirt exposed. Note: Use of band-style "oil filter wrench" compressors that do not lock to a set position is not recommended and can result in ring breakage!

18 Install Piston/Rod Assembly into Ring Compressor

After verifying ring clocking, lightly lubricate the skirt of the correct number piston/rod assembly and insert it into your sleeve-style ring compressor (or simply insert the piston skirt into the bore if using a band-style compressor). Start pushing the piston very lightly downward by hand. Further minor adjustments to adjustable-style ring compressors may be needed so that the piston can slide through it while still holding the rings firmly. When using a sleeve-style compressor, you will probably need to use your fingers and press each compression ring into its groove, to allow it to enter the tapered section of the sleeve (shown in left photo).

Important!

19 Avoid Rod Interference Problems

Once the piston skirt has entered the top of the bore and all rings are being compressed, look underneath to ensure the connecting rod is roughly centered and not about to contact any part of the crank or block (or another rod, if one has already been installed onto the journal). Shown is a rod hitting a crank counter weight, a common occurrence. Twist the piston as necessary to get the rod at a 90 degree angle to the crank centerline, and you may also need to slide the rod to center it along the piston pin. Failure to correct these problems can destroy the rod or other components!

20 Install Piston into Bore

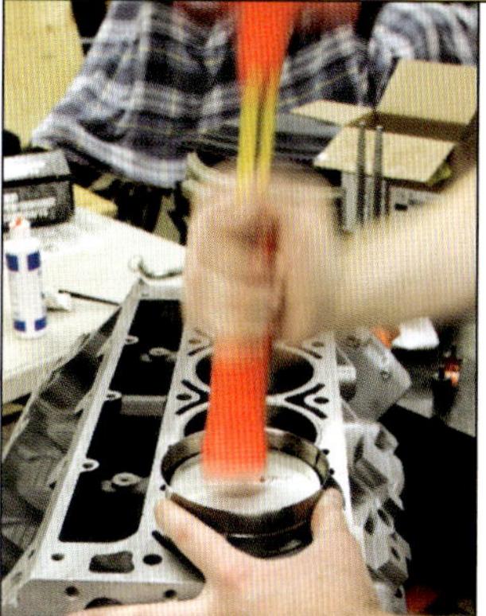

Hold the ring compressor against the deck surface and begin tapping downward on the top of the piston with the rubber butt end of a hammer. If the piston stops at any point, do not force it! The ring compressor may be improperly adjusted and a ring may be hanging up on the bore lip. Some moderately forceful taps may be needed to get the piston to move, but anything more than this should be a red flag—in which case you'll need to stop, pull the piston out, and make any necessary adjustments before trying again. Once you are certain the piston is going into the bore properly, reach underneath and begin guiding the end of the rod with one hand while continuing hammer taps with the other (you may remove the ring compressor at this point if you wish).

21 Guide Rod Onto Journal and Install Rod Cap

As the piston moves down, some further slight twisting of the rod may be needed to help guide it past the crankshaft counterweight (and other rod, as applicable). As the rod nears the crank journal, use extreme care, as any part of the rod touching the journal surface could cause a scratch. Hold your hand around the journal with your index finger and thumb, keeping the rod bolt holes centered on either side of the journal (shown, be prepared to get a little lube on your hand). Once the rod is seated onto the journal, install the rod cap (remember to do so in the correct orientation). On aftermarket rods that use locating dowels, it is recommended that you tap the cap in place using a rubber mallet (in lieu of drawing it down with the bolts) to seat the cap. Either way, install and tighten the bolts until snug. Repeat steps 15-21 until you have installed piston/rod assemblies into all cylinders.

Torque Fasteners

22 Tighten Connecting Rod Bolts

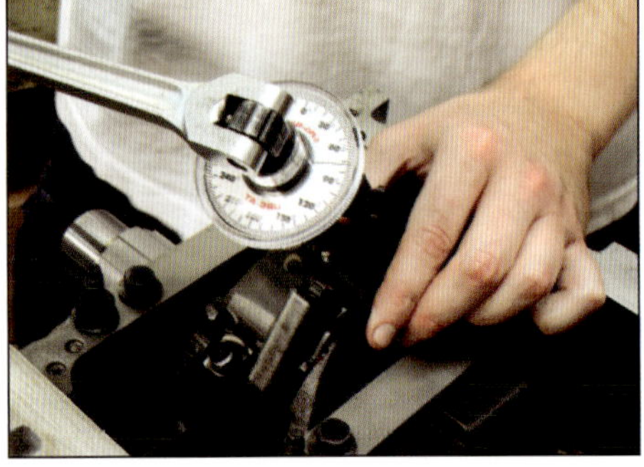

Note: The procedures described in this step apply to engines using GM connecting rods and rod bolts only. If using aftermarket rods or bolts, follow the tightening procedure specified by the manufacturer, and read "Utilizing Fastener Stretch" below for more information. *Flip the engine on its stand so that the oil pan rails are facing up. Starting with the #1 connecting rod, turn the crank so that both rod bolts are easy to access (the crank will be substantially more difficult to turn now that pistons and rings have been installed!). Torque each bolt initially to 15 ft-lbs. Then install your torque angle gauge and give each bolt an additional 75 degrees of twist. Mark on or near the bolt so you know you've tightened it, then ensure the rod can slide back and forth on the crank journal (if not, the rod bearings may be crooked). Repeat for each connecting rod until you have secured all sixteen rod bolts.*

Important Note:
Early-style GM rod bolts must only be twisted 60 degrees (instead of 75). The most foolproof way to determine whether your bolts are of this type is to look at the area between the bolt head and the threads: if the bolt has a thick shank with a series of shallow, vestigial threads starting just below the head, it is an early-style bolt. Later-style bolts all had a narrower shank interrupted by a larger-diameter sleeve somewhere between the bolt head and threads (exact location varied). The photo compares an earlier (left) and later-style (right) bolt removed from a rod, but you can have a look now just by loosening the bolts and pulling them out enough to look at the shank. Early- and later-style bolts should never be mixed on the same rod!

Utilizing Fastener Stretch

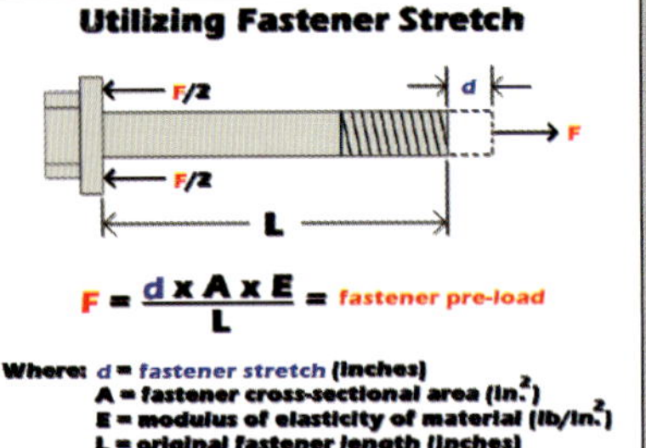

There are basically three options when it comes to measuring the tightness of a fastener, and the most common is the all-too-familiar torque specification. While this method gives acceptable results in most situations, the problem with a torque wrench reading is that it simply gives a measurement of how much twisting force is required to turn the head of a bolt (or nut). This can be influenced by factors like thread quality of the bolt and hole, and microscopic surface imperfections on the underside of the bolt's head. Though use of the appropriate lubricants (when specified) helps alleviate some of this inaccuracy, one can still never avoid the fact that torque doesn't always correlate with the amount of "pre-load" the fastener is experiencing (and hence, how much force it is actually clamping with). The torque-plus-angle method specified by GM helps compensate for some of these factors, but it's still not ideal.

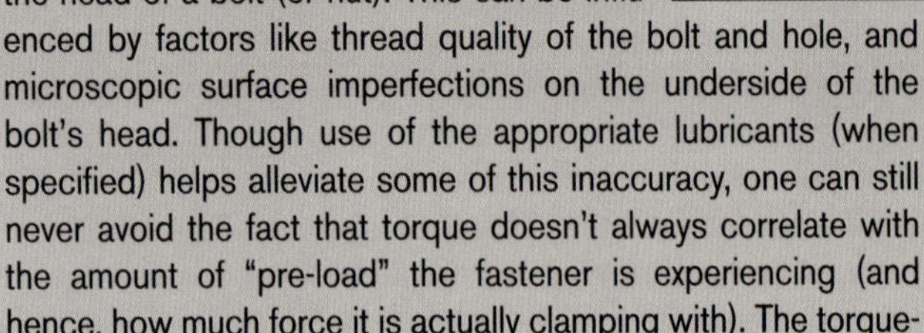

Rod bolts are some of the most highly stressed fasteners in an engine, and their pre-load must fall within a critical range: too little results in internal bolt forces that fluctuate and fatigue the bolt, and too much is harmful as well. For this reason, some connecting rod manufacturers use the so-called rod bolt stretch method, and while it can be time-consuming, it gives by far the most accurate reading of the clamping force provided by the bolt. So how does it work, you ask? As any bolt is tightened, it will stretch (elastically deform) slightly as its internal force progressively increases. It's exactly like stretching a rubber band, only the amount of stretch here is very small in comparison. By

Utilizing Fastener Stretch *CONTINUTED*

measuring this elastic deformation with a rod bolt stretch gauge and knowing the physical properties of the bolt, we get a direct indication of how much force it is holding the rod together with.

It is important to note that fastener stretch is not the same as tightening bolts "to yield"–one normally does not stretch rod bolts to the point where the stretch becomes permanent and they cannot be reused (as is the case with GM head bolts, for example). Read through the photo captions for hints on how to utilize the fastener stretch method to achieve the proper bolt pre-load in your high-performance connecting rods.

Before getting started, the threads of the bolt and the underside of its head are lubricated to prevent friction-related damage to the metal. You can see a dimple on the rod bolt head; there's another on its opposite end. These allow the rod bolt stretch gauge to grasp onto the bolt, and you will need to ensure that these areas (as well as the contact points on the gauge) are clean as a whistle; the bolt stretch being measured is very small and any dirt could ruin the readings. With the rod bolt just snug, the gauge is installed onto the bolt. These gauges are designed with spring-loaded action to hold themselves onto the bolt, and you want to be somewhere in the middle of this, so adjust as needed.

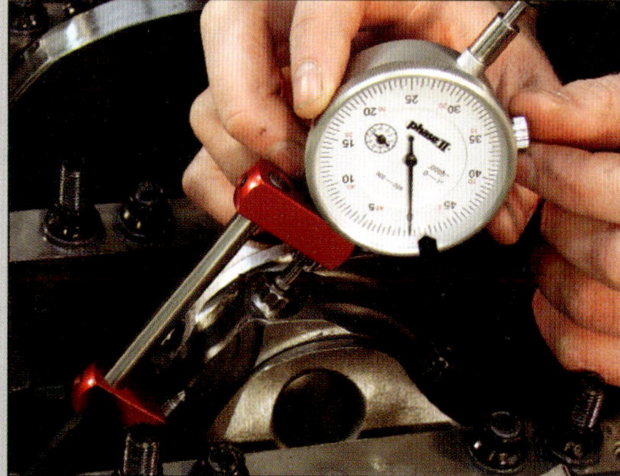

With the rod bolt stretch gauge installed, adjust its dial indicator to read zero. Take the gauge on and off of the bolt a few times, ensuring it returns to zero each time you are on the bolt. If it goes to different values, you likely do not have the gauge knobs adjusted properly or there is foreign material in the bolt dimples.

Ideally, the stretch gauge is left on while the bolt is turned with a box-end wrench, but the deep-skirt design of an LS block usually prevents such a wrench from reaching at least one side of the rod. You probably will need to remove the gauge, tighten the bolt a bit using a socket, and then put the gauge back on, repeating until you've got it right. Proper stretch values are normally in the range of around five thousandths of an inch, but be sure to go by the recommendation of the connecting rod and/or connecting rod bolt manufacturer.

23 Install Crankshaft Position Sensor

Being sure to use some oil on its O-ring (replace if damaged), install the crank sensor into the passenger side rear of the block. Torque its retaining bolt to 18 ft-lbs. Make sure you are using the correct sensor for your engine and computer: black sensors are for 24X reluctor rings, while lighter colored sensors are for 58X rings.

24 Short-Block Final Assembly Complete

The assembly of the short-block (or "bottom end") is now complete! Take a well-deserved break, but be sure to cover up your engine while you're gone to prevent contamination from airborne dust, bugs, and other unwelcome substances. Then, move on to the next step.

Professional Mechanic Tip

25 Install Camshaft

With the engine right-side-up on its stand, lubricate all cam bearings in the block that you can reach. Before continuing, be sure your hands are very clean, as it will be impossible to avoid contact with the camshaft's journal and lobe surfaces during installation. The front of the camshaft is recognized by its bolt hole(s) and sprocket locating pin. Lubricate the two rearmost cam bearing journals (and the lobes between them) with engine oil, then insert the cam until it can rest these journals on the front two cam bearings in the block. (Please note that there is no need for moly-based "cam break-in lube" on LS camshafts thanks to their rollerized design!) The fact that Gen III/IV small-blocks use five equally-spaced, equally-sized cam bearing journals allows insertion of the cam segment-by-segment like this, meaning you can lube the journals and lobes as you go along, making the process a bit cleaner. When well on its way in (and little remains exposed to grab onto), an LS cam's hollow construction lends itself nicely to the use of one or more long 3/8-inch extensions inserted into its central bore for added leverage (longer bolt(s) of the appropriate size and thread pitch can also be used, this is the only option for most single-bolt cams). Continue to insert the cam slowly and carefully (lest you mar the lobes or cam bearings), noting that more and more upward pressure will be required as the cam gets deeper into the block. When it is nearly all the way in, you can even insert a long 3/8-inch extension into the rear of the cam to help guide it the last few inches. Once the cam is in, remove the extension(s) or bolt(s), but be careful—the slight backward tilt of most engine stands means the cam may want to slide its way out the back of the block!

26 Install Camshaft Retainer Plate

Lightly lube the built-in gasket at the back of the cam retainer plate (check once more that it's in good condition). After wiping the corresponding surface on the front of the block, lubricate the thrust surface of the cam (its outside edge, which is inset slightly) and set the cam retainer in place. Use care not to accidentally push on the cam or it will slip backward and fall onto its lobes. Install the four cam retainer bolts and torque to 18 ft-lbs. If your cam retainer bolts are of a TORX-head design, the specification is 11 ft-lbs.

Special Tool, Professional Mechanic Tip

27 Install Crankshaft Key and Sprocket

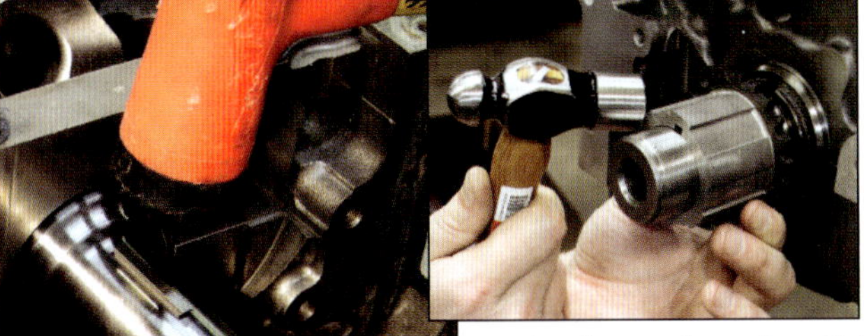

Install your crank key into the crankshaft snout's keyway using a rubber mallet. Be sure that it goes in squarely or it will be difficult to slide the crankshaft sprocket and/or oil pump drive gear over it. You may have installed your timing set's crank sprocket onto the crank snout during pre-assembly, but if not, do so now. Spray some lubricant such as WD-40 onto the snout to help this process. Some aftermarket sprockets will simply slide over the snout and only require a few light hammer taps to seat fully on the crank. Factory-style sprockets should be started with light hammer hits, but will need to be pressed on the rest of the way using a crank gear installing sleeve (possibly combined with a harmonic balancer installer tool). The alternative to buying a sleeve is to simply use your old crank sprocket (shown)! This method will also require your old crank bolt and is easier and cheaper than buying a special sleeve. Press the sprocket on until it firmly seats against the crank, then remove your old sprocket, it should just come off by hand.

Important Notes:

(1) Some cranks (expecially aftermarket ones) use a second key located further outward on the snout (to properly index aftermarket harmonic dampers and prevent any chance of it spinning on the snout). If using such components, install the second key at this time as well.

(2) If using an adjustable timing set, be sure to install the crank sprocket using the correct keyway you determined during pre-assembly.

(3) Do not use hard hammer hits to install the crank sprocket, this can cause severe engine component damage!

28 Install Camshaft Sprocket and Timing Chain

Turn the crankshaft until the alignment mark on its sprocket is at the 12:00 position. Spin the camshaft by hand until the cam locating pin is facing to the right (roughly 3:00), making sure not to push the cam backward at all. This pin placement will help get the cam sprocket locating mark roughly where it needs to be. Lubricate the thrust surface at the back of the cam sprocket and soak the timing chain in oil. Once this is done, take your cam sprocket and hang your timing chain on it. Reach behind the engine with one hand to hold the cam from moving backward (this also allows you to turn the cam slightly if needed). Put the cam sprocket in place while simultaneously wrapping the timing chain around the crank sprocket. You may have to take the cam sprocket and chain on and off of the engine a few times until you get the marks on the cam and crank sprockets to line up vertically (they should be at 6:00 and 12:00, respectively). Start the cam bolt(s) by hand (some aftermarket timing chains afford very little slack, making this difficult); torque to 18-26 ft-lbs (cams with three small retaining bolts) or 66 ft-lbs plus 40 degrees (cams with one large retaining bolt).

Important Notes:

(1) Be sure to use the correct sprocket markings determined during pre-assembly if using an aftermarket adjustable timing set.

(2) Engines equipped with VVT have the cam phaser mounted to the front of the cam sprocket, and the assembly is secured to the camshaft by an actuator solenoid valve in lieu of a bolt. It must be tightened to 48 ft-lbs plus 90 degrees.

(3) The timing chain tensioner (on engines so equipped) may need to be installed before the chain and cam sprocket. In this case, the tensioner will need to be temporarily deactivated using a pin or similar instrument while the chain and sprocket are put in place.

(4) Engines equipped with a timing chain dampener (which sits in the area between the cam and crank sprockets) should install it at this time, torquing its bolts to 18 ft-lbs.

29 Install Oil Pump

If your crank sprocket does not have the oil pump drive gear built-in, slide this item onto your crank snout now. Wipe the mating surfaces on the front of the block and back of the oil pump, then set the pump in place over the crank snout. You may have to twist the oil pump's gear teeth so that they align with the teeth on the crank sprocket (or separate oil pump drive gear). No gaskets or other sealants should be used between the pump and block surfaces. Torque spec on these bolts is 18 ft-lbs. Once the oil pump has been installed, flip the engine over on its stand.

Important Notes:

(1) If you are using an aftermarket double-roller timing chain, be sure to place any supplied spacers between the pump and block (and use any longer bolts supplied). Most aftermarket oil pumps do not require spacers as they are designed with the thickness of a double-roller chain in mind.

(2) While most pumps self-align properly to the block, LS7 oil pumps must be held flush or no more than 0.04-in. above the oil pan rails at the bottom of the block—no protrusion below the rails is acceptable! A straight edge or the GM J 41480 can be helpful in determining this—this is the same tool used to align the front and rear covers to the block (see steps 33-34). The procedure is similar for the LS9.

SLP Performance Parts

One aftermarket high-performance parts manufacturer you should have a look at for your rebuild project is SLP Performance Parts. SLP has been a major player on the LS engine scene ever since the LS1 first hit the streets in the late 1990s, and the company has developed an extensive line of internal engine components that includes the likes of rocker arms, ported cylinder heads, and high-lift camshafts.

In addition, SLP is also a great source for bolt-on external engine parts like exhaust headers and high-flow induction systems, both of which make a great complement to a newly-rebuilt, high-performance Gen III or IV. The company's Toms River, NJ facility is also an excellent place to take your LS-equipped ride for in-house custom ECM/PCM/TCM recalibration (see Chapter 9 for more information on tuning). For full details, direct your web browser to www.slponline.com.

SLP markets blueprinted, heavy-duty oil pumps with ported inlets and outlets for increased flow capability and reduced oil temperatures. This model also has an interchangeable pressure relief spring; when needed, the stiffer spring can be installed to afford about 5 psi of additional oil pressure, which is great for high-performance engines (especially those running wider-than-stock bearing clearances).

If you're putting together a stroker LS, you probably will want to get a set of SLP's oil deflector (a.k.a. windage) tray spacers. These spacers afford more clearance for the rotating assembly and will reduce or eliminate the need to modify the factory tray (as shown in step 30).

Performance Tip

30 Oil Deflector Tray Modification

Note: this step applies to high-performance applications only. For stock rebuilds, skip to the next step. *When using a crankshaft with a larger-than-stock stroke, modifications to your oil deflector tray may be required to obtain adequate rotating clearance. To check for contact between your rotating assembly and the tray, set it loosely in place atop the main bolts and rotate the crank, watching and listening for any interference. Most commonly, this occurs between the tray and the heads of the rod bolts. To correct for this, mark the tray at all points of interference, then take it off of the engine and use a hammer and chisel to bend the tray in these areas. Do not close off any gaps in the tray completely—this will create oil flow problems. Reinstall the tray and ensure you have adequate clearance, noting that some extra space will be necessary to account for crank stretch at high RPM! As an alternative to modifying the oil deflector tray, you may also install a set of aftermarket tray spacers (see "SLP Performance Parts" on page 122), but be aware of the following: because the oil pump pickup tube mounts atop the tray on most engines, some modifications (i.e. slight bending of its bracket) may be required in order to correct for decreased clearance between the pickup tube's screened inlet and the floor of the oil pan. This is easy to check with clay later during oil pan installation. A final note on oil deflector tray modification is that if you are using aftermarket main studs, they may require some of the holes in the tray to be enlarged slightly, which is easy enough to do with a drill or die grinder.*

31 Install Oil Deflector Tray and Pump Pickup Tube

Lay the oil deflector tray atop the main bolts, noting correct orientation (most are marked "REAR" at the back). It is a good idea to shoot some oil into the oil pump inlet at this point for initial lubrication. Install a new O-ring onto the end of the oil pump pickup tube. Coat the O-ring and the oil pump inlet opening with oil, then push the pickup tube into the oil pump. Ensure the tube is all the way in before inserting and tightening the retaining bolt, or else you may damage the O-ring or push it out of position. The retaining bolt gets 106 inch-lbs of torque, while the eight stock deflector tray nuts (one of which also secures the pickup tube bracket) receive 18 ft-lbs. If using aftermarket main studs and tray nuts (as shown here), use the torque specifications provided by the manufacturer.

Important notes:

(1) There are at least two different styles of pickup tube used on LS engines. Some tubes neck down near the end before bumping up to a flange. These tubes require a thicker O-ring, which is normally green in color. Other pickup tubes do not neck down and require a thinner (usually blue or black) O-ring; it is this type that is shown in the photos. An incorrect O-ring can cause loss of oil pressure and severe engine damage, so be sure you are using the correct type!

(2) If your oil pump has been spaced forward for use of an aftermarket double-roller timing chain, slight bending of the pickup tube bracket will be required for proper fitment.

(3) On dry sump engines (such as the LS7), the pickup tube is part of the oil pan, so only the deflector tray is installed at this time.

32 Install Crankshaft Oil Seals into Front/Rear Covers

The crankshaft's front and rear oil seals should not be reused, and must be removed from the engine covers and discarded. A hammer and flathead screwdriver can be used for this, though care must be taken not to score the aluminum surfaces of the covers. GM recommends waiting until the covers are on the engine to install new seals, but this requires special J-tools (which are invariably expensive or hard to get a hold of). The front seal is fairly easily tapped into place about its edges with a rubber mallet; do this slowly and gently or the seal will be destroyed. While the same can be done with the rear seal, you can also use this tool made by now-defunct Wheel to Wheel Powertrain (which tightens to squeeze the seal into place, second photo) to make the job more fail-safe. A thin film of oil applied to the engine cover surfaces will help the seals press into place. However, the seals themselves are designed to be installed dry—do not *lubricate their inner rubber surfaces! If you are afraid of botching this step, know that new front and rear covers are available from GMPP with seals pre-installed (see "GM Performance Parts" on page 125 for more information).*

Special Tool Used, Precision Measurement

33 Install Rear Cover

Before installing the rear cover, make sure the rear oil gallery plug (barbell restrictor) is still in place at the driver side rear of the block (see step 3)! Set a new gasket in place on the rear cover, using the first couple of threads of each rear cover bolt to hold it there. Wipe off, but do not lubricate, the crankshaft's rear flange. You must be very careful when sliding the rear cover onto the block, as it is easy for the lips of the rear crank seal to become misaligned while doing this, resulting in an oil leak (updates to GM's seal design have made this much more foolproof, though). The aforementioned Wheel to Wheel Powertrain tool's aluminum "donut" helps ease an earlier-style seal's transition onto the crank. Once all rear cover bolts are started by hand, you have a choice. The first option is to use a GM cover alignment tool (J 41480) to align the rear cover's cover-to-pan sealing surface with the block's oil pan rails before torquing the rear cover bolts to 18 ft-lbs. As an alternative, you may visually align the cover-to-pan sealing surface with the pan rails, tighten the bolts, and then verify no more than a 0.020-inch drop between the pan rails and rear cover using a straight edge and feeler gauge (any protrusion of the cover beyond the pan rails is unacceptable). This latter method will usually provide acceptable results since contact between the rear seal and crank flange helps roughly align the cover to the block.

GM Performance Parts

One of the best sources for both high-performance and stock-replacement LS engine parts alike is GM Performance Parts. The company has always been a great product source for the original small-block Chevrolet, and with the advent of the LS, the tradition of excellence and innovation continues. In fact, with this GM branch regularly unleashing new parts for Gen III/IV engines, it's often hard to keep up with the latest offerings!

Simply put, GMPP's line of OEM and OEM-quality products is about as good as it gets when you're talking about components for the LS engine family. We highlighted some of the company's parts back in Chapter 4, but wanted to clue you in on a few others that may be of interest during your rebuild project; check out the photos and captions here for the low-down. From camshafts to high-strength engine blocks, and from engine covers to fasteners, GMPP literally has it all! For more information or to get a copy of the latest catalog, check out www.gmperformanceparts.com.

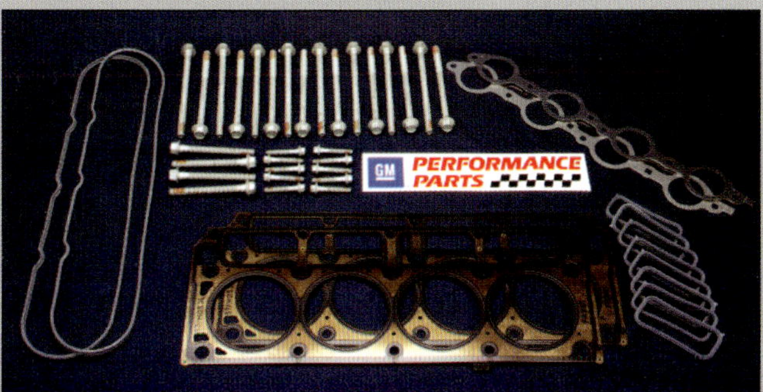

Though obviously designed with home hot-rodders performing head/cam swaps in mind, GMPP's Cylinder Head Installation Kits lend themselves well to use in stock and high-performance rebuilds, too. These kits (which are available to fit many LS engines) include new head bolts, MLS head gaskets, intake port seals, exhaust manifold gaskets, valve cover gaskets, and other items depending on application—all packaged together and ready to go.

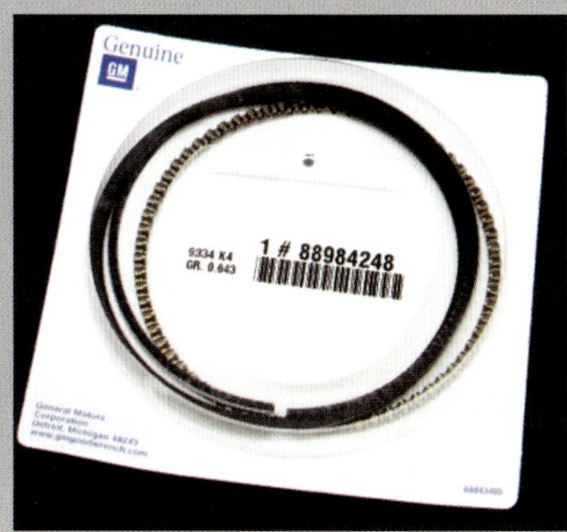

We showed you a stock-replacement GMPP piston back in Chapter 4, but what's a piston without the rings to match? Happily, GM Performance Parts offers an array of LS piston ring kits in both stock and oversize form. This is a set of 0.010-inch oversize rings for LS1/LS6 engines.

Building an LS engine from scratch? GMPP packages most everything you need to top off your block with a Bare Block Completion Kit. Available in different configurations based on block generation, this is also a great way to replace old engine covers and hardware that may have suffered from the elements and aren't quite as shiny as you'd like them to be. The covers also include new seals pre-installed, which is a boon for those afraid of botching step 32! (Please note that parts included may vary from those shown.)

Swapping your fresh LS into a different vehicle than it was originally installed? GMPP sells the oil pan you need to clear your K-member or other undercar chassis components. This pan is the same one used on 1998–2002 Camaros and Firebirds.

Special Tool, Precision Measurement  

34 Install Front Cover

The front cover goes on next, being sure to use a new gasket behind it. This cover is most easily aligned using the same alignment tool used for the rear cover, along with an additional J 41476 tool to help align the cover side-to-side. (This latter tool was not used in the previous step since the rear seal had already been installed and was basically serving the same purpose). Install the J 41476 hand-tight using your old crank bolt before installing the J 41480. If you do not have access to these two tools, it is strongly recommended that you wait until after you install the harmonic damper (step 59) to tighten the front cover bolts. This will help align the cover side-to-side and will of course require the oil pan to be installed after the harmonic damper as well. Either way, the front cover bolts get 18 ft-lbs of torque, and you must also verify no more than a 0.020-inch drop between the pain rails and the cover-to-pan sealing surface.

Important notes:

(1) Some aftermarket oil pumps require grinding to the inside surfaces of the front cover for proper fitment. Be sure to clean all shavings from the cover and seal after doing this.

(2) The LS7 does not use the J 41476, but rather its own specialized set of cover alignment tools, which require that the front crank seal not be installed until afterward. Absent access to these tools, it is recommended that you take the route of installing the harmonic damper first to help align the LS7's front cover. The LS9 installation is similar.

(3) Engines equipped with VVT have provisions for actuating the cam phaser mounted to the front cover (you should have noted their layout during disassembly), so be sure to install these components now as well.

35 Prepare to Install Oil Pan

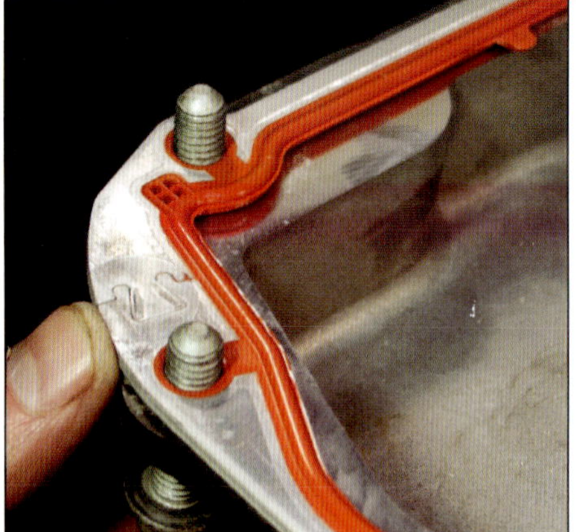

Oil pan design varies substantially by engine and application, however all are installed using the same methodology. First, make sure any and all internal baffles, the oil filter adapter, the oil filter bypass valve, and like items that you may have removed during cleaning are in place and tight. Then set your new pan gasket atop the oil pan, using as many bolts as you can to hold it (it is not necessary to actually rivet the gasket to the pan, though you can if you wish). Now, apply a 1/4-inch bead of RTV silicone at each of the four meeting points of the block and front and rear covers. You will note that the front and rear cover gaskets protrude toward the pan gasket slightly at each of these points, but this RTV is for extra insurance.

36 Install Oil Pan

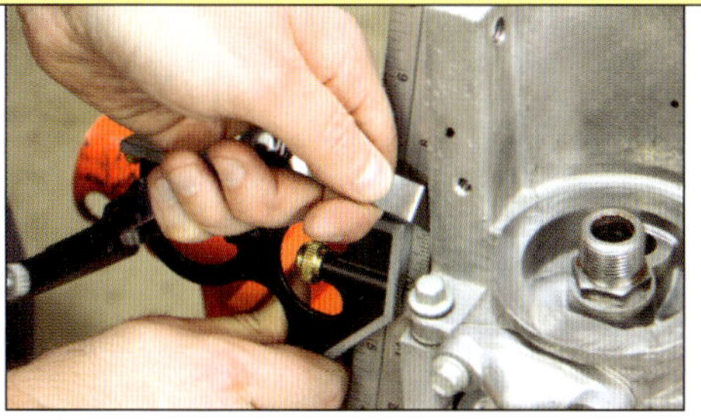

Now set the oil pan in place on the engine block. Depending on the baffle design of your pan, it may not drop straight down, but rather require some wiggling and/or angling to install. A properly installed pan will sit flush on the engine block; if it does not, your pickup tube may be hung up in the oil pan baffle(s), so lift up and try again. Once you are confident the pan has seated correctly, tighten the bolts only snug *and use a straight edge to measure the pan's location with respect to the rear of the engine block. Because the oil pan forms a structural part of the driveline (heck, even some of the bellhousing bolts attach to it), this dimension is critical: the pan cannot protrude beyond the back of the engine block, and may only be set forward 0.010-inch on most engines. On some engines, a maximum of just 0.004-inch is considered acceptable! After making any necessary adjustments to get the pan location correct, tighten all the short oil pan bolts to 18 ft-lbs. and the two long pan-to-rear-cover bolts to 106 in-lbs (most engines; pan and bolt style may vary). After verifying pan alignment is still acceptable, replace any and all sensors you previously removed from the oil pan (for example, the oil level sensor and oil temperature sensor, these vary by application).*

37 Install Camshaft Position Sensor

Turn the engine right-side-up on its stand. On Gen III engines, the cam sensor slides into the top rear of the block, and its retaining bolt is tightened to 18 ft-lbs. On Gen IV engines, the sensor is installed into the front cover, and its bolt gets 106 in-lbs. (the sensor on VVT-equipped engines is similar but not identical). With either style, be sure to use some oil on the O-ring. Many Gen IV engines also have a wiring harness extension and bracket (leading to the bottom of the cover) that you may wish to install now.

38 Install Valve Lifters and Guide Trays

The best way to install the lifters into the engine is to first insert them into the lifter guide tray. Apply some engine oil to the grasping areas of the tray, then slide the lifters in. Because the tray grabs onto the flat areas on either side of the lifter, the lifters will only go in one of two ways, with either being acceptable (orientation of the lifter's oil hole on the side does not matter). Then spread oil on all surfaces of each lifter—roller tip included—and push the tray into place in the engine. Each lifter should slide easily into its bore. On some engines, the shape of the tray dictates that it can only be installed in one orientation, but on others this does not matter. Tighten the lifter guide's retaining bolt to 89-106 in-lbs—do not overtighten and crack the tray! Repeat for the other three trays until you have installed all 16 lifters.

Important Notes:

(1) It is recommended that you not soak hydraulic lifters in oil before installation, as this can interfere with proper rocker tightening (stock rebuilds) or with valve lash adjustment (high-performance rebuilds using adjustable rocker arms).

(2) On AFM-equipped engines, the lifters for cylinders 1, 4, 6, and 7 look slightly different, mainly in that they have built-in springs that allow them to follow their cam lobe profiles while deactivated. Be sure to install them into their correct locations. The areas that the guide tray grabs onto are also of a different shape on these cylinders.

39 Press in Cylinder Head Locating Dowel Pins

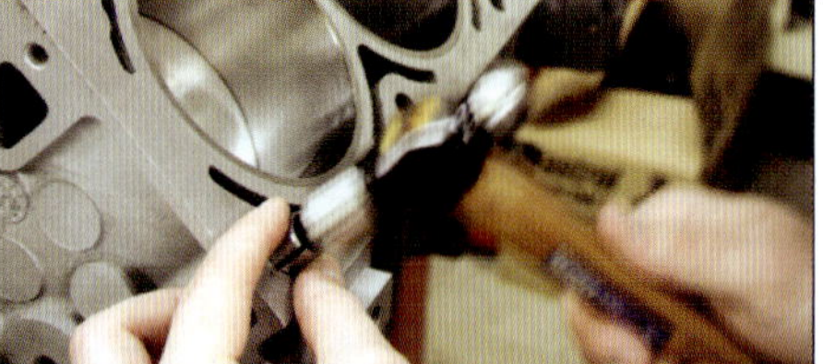

There are two dowel pins for each deck surface of the block, one at the front and one at the rear. It's best to not reuse your old dowels, so install new ones now. Each must be pressed into its hole as completely as possible, and light hammer taps may be needed to do this.

40 Lay Head Gasket on Block

Wipe the deck surfaces of the block clean one last time before proceeding. Tilt the engine in its stand so that one deck surface is horizontal. Take a head gasket and place it on the block, noting any markings on the gasket such as "FRONT" or "THIS SIDE UP." Some gaskets are unmarked; to determine their orientation, note that because the rear portion of many LS blocks has more coolant passages between the block and heads, these must match any corresponding holes in the head gasket. Also, you may have to press down on the lower corners of the gasket to seat it onto the dowel pins.

Critical Inspection

41 Inspect Cylinder Head for Gasket Incompatibility Recess

Some early Gen III heads have a recessed area along the edge of the deck surface, just below the number 3 or 6 exhaust port (the head shown does not have the recess, but it would be exactly in the area pointed at). If your head has a recess, you cannot use GM's newer-style MLS head gaskets, but must rather use a GM graphite-layered steel core gasket. See Chapter 4 for a comparison photo of these two types of gaskets. In the case of aftermarket gaskets, consult with the manufacturer for head compatibility information. If your heads are not currently assembled, follow "Cylinder Head Assembly" Workbench Tip before proceeding to step 42.

42 Install Cylinder Head and Bolts

Make sure that the deck surface of the head is completely clean. Grasp a head along its sides (holding fingers inside the intake ports works well) and set it in place atop the head gasket. Ensure that it locates properly on the dowel pins. There are fifteen head bolts: ten that are large in diameter (M11 thread on most engines) and five small in diameter (M8 thread). Insert them at this time and turn them until they are just barely snug against the head.

Important notes:

(1) If you have a 2003 or earlier block, two of the M11 bolts are shorter than the rest, and these must be installed at either end of the top row of M11 bolts (locations #9 and #10 in the diagram shown in the next step; there is one shown being inserted in the photo to the left). On 2004 and later blocks, all M11 bolts are of the same length.

(2) While you must always install new M11 bolts, your old M8 bolts can be reused so long as you apply medium-strength threadlocker to the first several threads.

Cylinder Head Assembly

Unless you've purchased a set of complete high-performance cylinder heads (that came with valvesprings and related hardware already installed), or have had your machine shop assemble your reconditioned heads for you, chances are that you need to perform cylinder head assembly now.

Here are a couple of important tips to keep in mind during this process. First, if your machine shop uniquely matched each valve to a given location in the head, you will need to keep close track of this and install all valves to their correct positions. Also, you will need to have any head-related notes you took during pre-assembly handy, as you should have written down the required location of any shim(s) needed to adjust valvespring installed height (see Chapter 7, Step 26).

1 Ready to Begin Cylinder Head Assembly

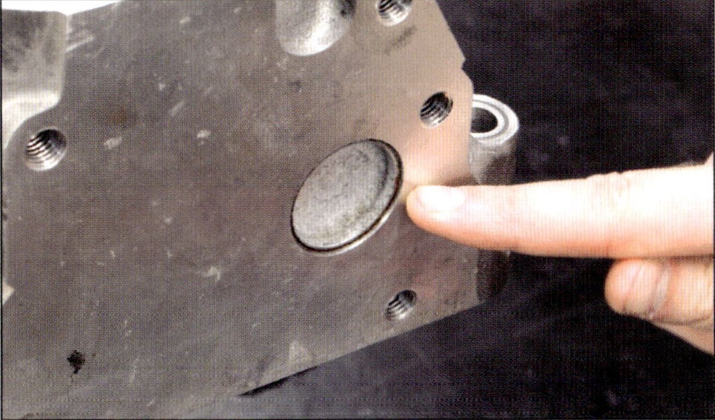

As during disassembly, it's best to have some cylinder head inspection stands to ease the head assembly process. Before beginning, look at both ends of the head to make sure any freeze plug(s) are still intact. If not, press these plugs in now using some medium-strength threadlocker around their edges (note that not all LS heads have provisions for such plugs).

2 Install Valve Stem Oil Seals and Spring Seats

Most LS valvestem seals are a one-piece design that includes the valvespring seat. These install easily: use an appropriate size deep socket and a hammer, tapping gently and straight down until the seat is firm against the surface of the head. Seals are often different for the intake and exhaust valves, as is the case with these seals acquired from GMPP (black are for the intake valves, red are for the exhaust valves). Please note that a special tool is required to properly install the seals of most types of aftermarket (and earlier-style GM) 2-piece seals/spring seats (the seat serves to properly align the seal on 1-piece units), so in this case we recommend having your machine shop perform this step for you.

3 Install Valve

Starting at one end of the head, lubricate the interior of a valve guide with engine oil. Also apply oil to the stem of the appropriate valve and slide it into place from underneath the head. Spin the valve as you insert it to spread oil evenly throughout the guide, and use care to move slowly as the tip and valve lock grooves pass through the valve seal. If the head is sitting horizontal, you may need to hold the valve from below to keep it from falling for the time being, though the tension of most seals will hold the valve in place. (If not, you may wish to perform this process with the head standing on its side or resting on the intake manifold flange.)

Cylinder Head Assembly *CONTINUTED*

4 Install and Compress Valve Spring

Set any shim(s) needed in place atop the valve spring seat, then place the valvespring atop them, followed by the valvespring retainer. Put on some safety goggles and use your valvespring compressor tool to compress the valvespring, being sure the lower end of the tool is centered on the valve. As during disassembly, make any adjustments needed to the compressor's jaws to compensate for retainer width, and again, the bottom portions of most tools have a knob that is turned to change the height at which the tool locks in place once the valvespring is compressed. You will want the tool adjusted such that you have just enough room to insert the locks. If the spring is compressed too far, they'll simply fall out of their grooves.

5 Install Valve Locks and Remove Valvespring Compressor

With your safety goggles still on, insert your valve locks in place about the center of the retainer– this may take some angling of the compressor tool to allow sufficient space for each lock to slide between the stem and retainer. It works best if you push one lock into a narrow area of the gap first (this will roughly center the valvestem), then insert the other lock. Then slowly remove the valvespring compressor, ensuring that the valve locks are seated properly once the valve is closed and the tool removed (some amount of gap between the locks is normal). You may gently tap the top of each retainer with a hammer to check this. Repeat steps 3-5 for all valves in the head, and also install wear/lash caps to the tips of the valvestems if using titanium valves.

6 Cylinder Head Assembly Complete

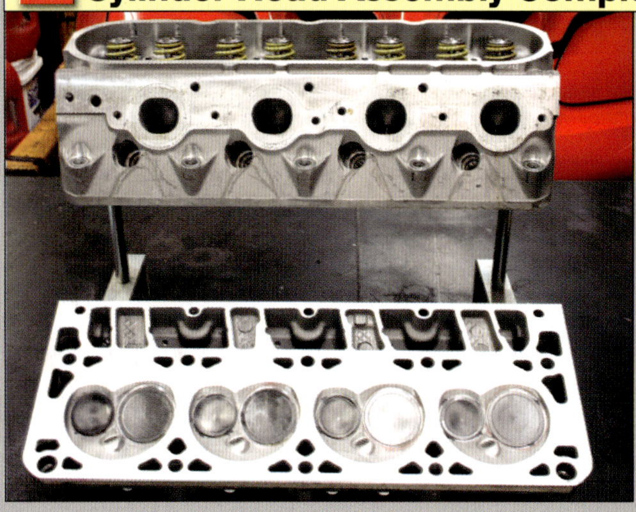

After repeating steps 1 through 5 for the other head, cylinder head assembly is complete. At this point, you may wish to install the coolant temperature sensor into what will be the driver side head (this is the usual configuration for LS engines), as well as the plug removed during disassembly into what will be the passenger side head. (Or, you may wait until later on to do this, there may be a risk of damage to the sensor during installation of the engine into your vehicle, depending on application.) Use Teflon-based sealant on the threads and torque each to 15 ft-lbs (not shown).

Torque Fasteners 

43 Tighten M11 Head Bolts

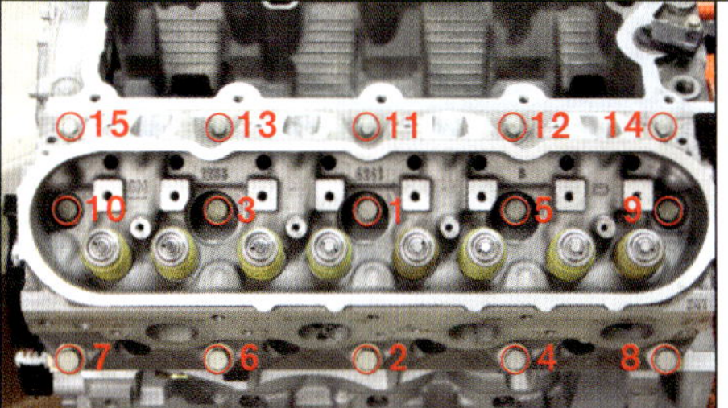

Refer to the accompanying image for bolt numbering. The M11 bolts must be tightened in three steps. These bolts require a substantial amount of force to twist, so you may wish to have an assistant help you hold the engine stand from rolling around while you're doing this!

FIRST STAGE: Using a torque wrench, tighten the M11 bolts numbered 1 through 10 in sequence to 22 ft-lbs.

SECOND STAGE: Now use a torque angle gauge to add 90 degrees of twist to these same bolts, again in sequence.

FINAL STAGE: The specifications to use in this stage depend on the style of head bolts that your block uses.

2004 and later blocks with all-same-length M11 bolts: add an additional 70 degrees of twist to the bolts in sequence.

2003 and earlier blocks with long/short M11 bolts: add an additional 90 degrees of twist to the bolts numbered 1-8 in sequence. 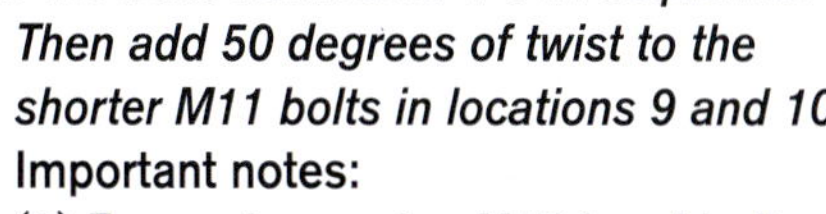
Then add 50 degrees of twist to the shorter M11 bolts in locations 9 and 10.

Important notes:

(1) For engines using M12 head bolts (such as the LS9), refer to your GM service manual.

(2) If using aftermarket head bolts or studs, refer to the installation instructions and tightening specifications of the manufacturer. However, the above tightening sequence should always be used.

Torque Fasteners

44 Tighten M8 Head Bolts

Once all M11 bolts are secured, use a torque wrench to tighten the M8 bolts numbered 11 through 15 in sequence to 22 ft-lbs. There is no angle to add to these bolts. With all head bolts fully tight, repeat steps 40-44 for the other cylinder head.

Professional Mechanic Tip

45 Mask Off Intake Ports

As the valves are about to become operational, you do not want stray bolts or other items accidentally entering the cylinders, requiring head removal. Avoid any chance of this by masking off the opening of each port–especially those of the intakes–with tape. Once applied, you may want to punch a small hole into each piece so that the engine can "breathe" as rockers are being installed and the crank is turned over (this latter concern doesn't apply so long as you have not yet installed spark plugs).

46 Install Pushrods

Lubricate the lower tips of all sixteen pushrods with engine oil, then slide each into place through its passage in the cylinder head. Press down on each one as it is inserted to make sure the tip is centered onto the lifter plunger, and also to push the lifter down onto the cam. For the moment, do not lubricate the upper tips of the pushrods.

47 Prepare to Install Rocker Arms

On engines using stock or small-duration aftermarket camshafts, it's possible to install and tighten the rocker arm bolts eight at a time at just two different positions of crankshaft rotation. However, since long-duration cams can cause some lobes to command lift at these positions, we're going to go through a procedure that will work regardless of camshaft used. First things first, though: thanks to the rollerized design of factory LS rockers, we suggest submersing them in engine oil to help lube the internal bearings. After this, you must lubricate each tip with

Important!

47 Prepare to Install Rocker Arms *CONTINUED*

engine oil and liberally lubricate the pushrod cup (a specialized assembly lubricant is strongly preferred over oil for this area because of the amount of friction that will occur here at startup). Also lube the area that will contact the underside of the rocker bolt head with oil to assist in proper tightening, and use your oiling can to squirt plenty of oil between the valve spring coils onto the valvestems. Then, for cylinder heads where the rocker pivot supports are not a machined part of the head (this is the case with most LS heads), lay the rocker stands in place on each head; some aftermarket stands get bolted in place. Please note: *because our method of rocker installation does not involve installing all rockers simultaneously, the factory stands* must *have at least 2 rocker bolts inserted into each of them at all times (spaced as far apart as possible) in order to prevent the stands from shifting out of place—the built-in factory alignment tabs alone will* not *stop this from happening!*

Professional Mechanic Tip

PRO TIP

48 Install Rocker Arms

1

2

3

4

5

6

PRO TIP Note: this step applies if using stock-style rocker arms only. If using aftermarket adjustable rocker arms, skip to step 49. *The rocker installation process will require some attention—along with an assistant—because of the tendency of the lifter guide trays to hold the lifters off of the cam once they are raised, which masks lifter movement. However, when done correctly, the results will be properly tightened rocker mounting bolts on any application! 1. Start off by having your assistant use your old crank bolt to turn the crankshaft clockwise (most crank turning tools can no longer be used thanks to the front cover being in the way). 2. Hold downward tension on the pushrods for cylinder #1. This prevents the lifter guide trays from holding the lifters up artificially. Wait until the exhaust pushrod* just begins *to move upward (the exhaust is the one on the* right *when looking from the side of the engine). 3. At this point, lube the pushrod tip and install the intake rocker arm (make sure the pushrod enters its cup on the underside of the rocker properly) and torque its mounting bolt to 22 ft-lbs. 4. Continuing to hold tension on the exhaust pushrod, have your assistant turn the crank until your newly installed intake rocker rotates the intake valve open and then almost completely closed. Now install the exhaust rocker arm, torquing its mounting bolt to the same 22 ft-lbs. 5. Repeat this step for every cylinder until all sixteen rockers have been installed.*

Important notes:

(1) Gen IV-style heads use rockers with offset tips for the intake valves. If you have this type of head, remember to install these rockers in the correct locations. (Recall, though, not all Gen IV engines use "Gen IV-style" heads, see our "LS Cylinder Head Evolution" Workbench Tip in Chapter 4 for more information.)

(2) If using an aftermarket cylinder head where the intake rocker mounting bolt intersects the intake port, it is not a bad idea to apply some RTV sealant to its threads.

(3) If you're getting confused about which lifter is on the base circle of the cam at any given point, it helps to verify by viewing the lifters through the cast-in "gaps" at the front and rear of the valley. See above, picture 6. Unfortunately, this is only possible for cylinders 1, 2, 7, and 8 since these lifters are the only ones visible through these gaps. The alternative is to wait until after rocker installation to install the engine covers; this way, the cam sprocket will be visible during this step to help you sort things out.

Performance Tip 

49 Installing Aftermarket Adjustable Rocker Systems

Note: this step applies to high-performance applications only. For stock rebuilds, skip to the next step. *When using aftermarket adjustable rocker arms, the installation procedure is often very similar to that discussed in step 48, and you should always follow the instructions provided by the manufacturer when using such a system. However, here are a few helpful hints to guide you if you're using a* shaft-mount *rocker system (this is not the only style of system available for LS engines, but they are by far the best). With installing an adjustable shaft rocker system, the main thing to keep in mind is to back the lash adjusters off all the way (i.e., as high as possible) so that they do not interfere with the process of tightening the mounting bolts. Also when installing and tightening their mounting bolts, since most shaft rockers are joined together as one pair for each cylinder, both lifters must simultaneously be on the base circle of the cam lobes (rotate the crank until a bit after the intake lifter returns to its lowest point). Once a rocker pair is installed, each rocker is then individually adjusted as follows: the lash adjuster is turned until it just contacts the tip of the pushrod ("zero lash"), then a specified number of turns (usually 1.5 to 2) are added to achieve the proper lifter preload. The lash adjuster is then locked in place via a nut.*

50 Install Valve Covers

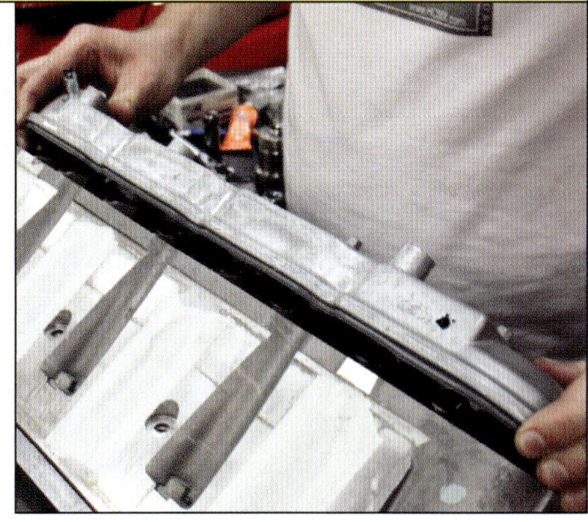

With all rockers installed and adjusted (as applicable), we can button up the top of the cylinder heads with the valve covers. Most LS cylinder heads use four centrally located bolts to hold the valve cover in place. These bolts must be torqued to 106 in-lbs in an inner-to-outer sequence. New valve cover gaskets are mandatory, though the grommets under each of the bolts are OK to reuse.

51 Install Valley Cover or LOMA

Set the valley cover (or LOMA on AFM-equipped engines) atop the block with a new gasket in place. Some aftermarket intake manifolds require substitution of the stock valley cover bolts, as their heads are too tall to afford adequate clearance from the floor of the intake. Either way, torque these bolts to 18 ft-lbs.

52 Install Oil Pressure Sensor

On Gen III engines, install the oil pressure sensor at the upper rear of the block (next to the cam sensor, shown); tighten to 15 ft-lbs. On Gen IV engines, the oil pressure sensor installs into the rear of the valley cover and receives 26 ft-lbs of torque. Be sure to reuse the sealing washer under the sensor (as applicable) and use some Teflon-based sealant on the threads (new sensors normally already have sealant applied).

53 Knock Sensor Installation

If your engine is a Gen III, the knock sensors mount into the top of block, in the valley cover area. Carefully lower each sensor into place and torque to 15 ft-lbs. If your engine is a Gen IV, the knock sensors install low on either side of block (not shown); the retaining bolt of each gets 15-18 ft-lbs.

54 Installing Knock Sensor Wiring Harness

Note: This step applies to Gen III engines only. For Gen IV engines, continue to the next step. *Because the intake manifold is about to obstruct access to the top of the valley cover, the piece of wiring harness that connects to the knock sensors on Gen III engines must now be plugged in. Run the wiring to the rear of the engine, and make sure the seals seat properly in the valley cover as well.*

55 Install Coolant Air Bleed Pipes

*The last items to be installed before the intake manifold are the coolant air bleed pipes (or blockoff caps, if originally equipped at the rear—*never *install blockoff caps between the fronts of the cylinder heads!). As discussed when they were removed during disassembly, exact style of these pipes varies by engine; however, all are installed with bolts tightened to 106 in-lbs, making sure a gasket or O-ring is used in between.*

Torque Fasteners

56 Install Intake Manifold

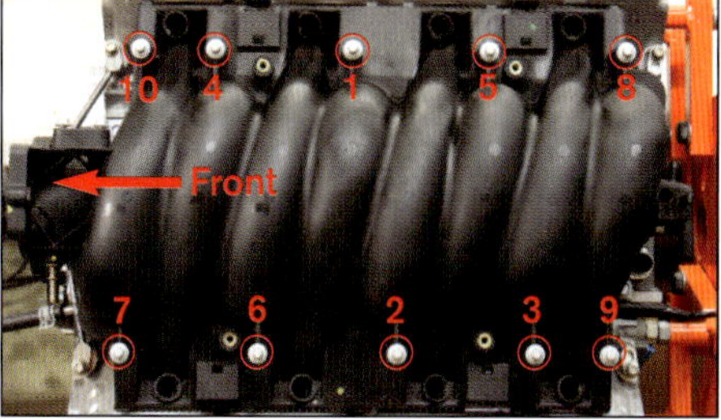

Before setting the intake manifold in place, install new intake port seals (car intakes, shown) or clip new carrier-style gaskets (truck intakes) in place. The intake port seals may want to creep out of their grooves, so ensure they are seated correctly before proceeding. Remove the masking tape from your intake ports and wipe the sealing surface of the heads, then place the manifold on the engine. Start its ten bolts by hand (use some threadlocker on their threads). You will need to make sure any brackets that attach to the intake manifold (and were removed during disassembly) are returned to the same place before doing this, as some are held by the intake bolts. Torque the bolts, in the sequence shown, a first pass to 44 in-lbs and a second pass to 89 in-lbs. (Note: For factory-supercharged engines such as the LS9 and LSA, refer to your GM service manual for installation instructions.)

57 Install Fuel Rails, Throttle Body and Intake Manifold Accoutrements

Exact style of fuel rails and injectors varies by application, but they should be installed now using threadlocker on the bolts. Fuel injector O-ring seals should be lightly lubricated with engine oil to ease insertion, and again, if reusing your injectors, use of new seals is mandatory! Throttle bodies are installed via either 3 or 4 bolts, and a new seal should be used for this, too. Now is also a good time to install the MAP sensor, any PCV system hoses, brackets, throttle body sensor(s), and any other items originally removed during disassembly. The detailed notes and photos you took regarding these items will pay off big time here!

Torque Fasteners, Special Tool

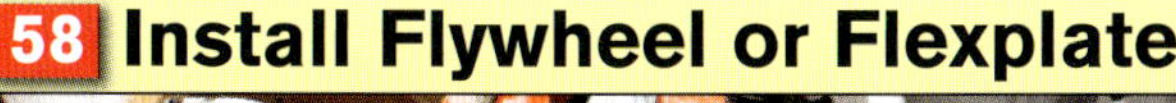

58 Install Flywheel or Flexplate

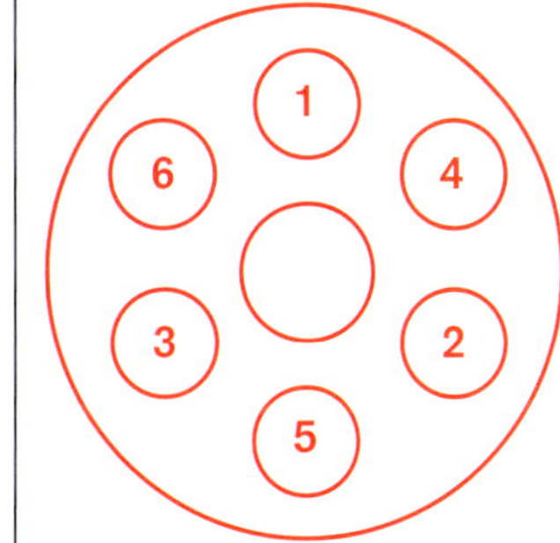

Line up the "extra" hole in the flywheel or flexplate with that in the back of the crankshaft (if applicable). Apply threadlocker to the threads of the six bolts and install them. Torque to 15 ft-lbs using the sequence shown in the accompanying diagram, then repeat this sequence for additional stages of 37 and, finally, 74 ft-lbs. A flywheel holding tool is suggested for this step, though you can also have an assistant insert a pry bar into the flywheel/flexplate's teeth (as may have been done during disassembly). Note that some ultra-high-performance engines such as the LSA and LS9 use an 8- or 9-bolt crankshaft flange; in this case, refer to your GM service manual.

Professional Machanic Tip, Special Tool 

59 Install Harmonic Damper

Because of the high bolt torques needed for proper harmonic damper ("balancer") installation, it is strongly recommended that you now install a flywheel holding tool to keep the crankshaft from turning over. There are two methods that can be used to press an LS damper onto the crank snout. The first involves using a harmonic balancer installation tool, which is fairly self-explanatory (not shown). But back in Chapter 2, we also mentioned that an alternate method is available, and here it is: acquire an M16 x 2.0 x 120mm bolt along with at least a few washers. Use this longer bolt (with the washers underneath to help it spin) to pull the damper onto the crank snout, switching to your old crank bolt if the longer one bottoms out. Watch to ensure the crank seal in the front cover stays intact during this process (do not lubricate the seal or the corresponding surface on the damper). With either method, once the balancer is on the crank as far as it will go, install your old *crank bolt and torque to 240 ft-lbs. Remove it and ensure the crank snout is recessed no more than about 0.175-inch from the innermost ridge of the balancer. Only then may you install your* new *crank bolt, torquing to 37 ft-lbs plus 140 degrees of twist.*

Important notes:

(1) Never attempt to press on the damper using just your old crank bolt! You will destroy the first few threads in the crank snout.

(2) Some engines use a special locking washer between the front of the crank snout and the balancer (you should have noted this during engine disassembly). Be sure to install it during this step, as applicable.

(3) Because of the different length crank snouts used in some Gen IV engines (for example, LS4 and LS7), a different length bolt may be needed than that listed.

(4) While most factory cranks and dampers are not keyed to one another, many aftermarket ones are. In this case, you will need to align the keyway in the damper with the key in the crank for proper installation.

60 Final Engine Assembly Complete

At this point, final engine assembly is roughly complete. We've installed every key engine component, and basically everything that's left is better considered an accessory! Your stage of desired completion at this time may vary; you'll need to think back to engine removal and disassembly to decide "how much further to go" while the engine is still on its stand, since further component installations may hinder your ability to easily install the engine into the vehicle and/or onto the vehicle's removable subframe. Some of the items you may wish to bolt up now, but which we will not go through in detail, include:

- *Starter motor*
- *Engine mount brackets*
- *Spark plugs*
- *Ignition coils and brackets*
- *Water pump*
- *Other accessory brackets*
- *Exhaust manifolds (use threadlocker on these bolts)*
- *Oil filter*
- *AIR system*
- *Dipstick tube*
- *Clutch pilot bearing/bushing*

Most of these items are pretty self-explanatory to reinstall: just reverse the steps you went through during disassembly (see the Appendix for some helpful torque specifications). That said, move on to Chapter 9 for hints on installing, tuning, and breaking-in your newly-rebuilt LS!

Break-in and Tuning

Congratulations on a job well done—you've got a newly rebuilt LS sitting on an engine stand! Though the work of this book is basically over, we wanted to leave you with some final information that will help ensure your engine enjoys a trouble-free reinstallation and startup, followed by a long useful life.

Engine Reinstallation

We finished Chapter 8 on an open-ended note, stating that you would need to reinstall all remaining engine accessories (like the exhaust manifolds and water pump) either while the engine was still on its stand or at the appropriate time during reinstallation. This may be easiest with the engine on its removable subframe (if applicable), or you may have to wait until the engine is already under the hood.

As with engine removal, reinstallation of your LS engine into your vehicle is beyond the scope of this book. There isn't much to say here except to use patience, take all appropriate safety precautions, and refer to any photos and notes you took during engine removal to sort out what goes where. Use all fastener torque specifications specified in your GM service manual and have someone else helping you throughout the process—engines and vehicles (along with their chassis components) are heavy and can do serious damage if not handled correctly!

Reinstallation of your engine will take some time and care, but one bit of good news is that it shouldn't be nearly as messy as during removal, since you won't be draining any fluids! In addition to the wiring harness, exhaust piping, cooling system hoses, fuel lines, emissions hoses, and all other engine-related systems, be sure to reconnect any brake, power steering, and A/C lines you may have disconnected during engine removal.

Final Preparations

With your engine successfully reinstalled into the vehicle, it's time to stop and double-check your work. Carefully examine every inch of the engine bay to make sure you haven't missed any electrical plugs or forgotten to reconnect any hoses. Once you're certain everything checks out, use factory-specified procedures to replenish and bleed all fluids (as

Although there are many post-installation checks to make that are important for your personal safety, the safety of your engine requires that you fill your crankcase with the correct amount, and type, of engine oil.

necessary). These vary by vehicle, but the two fluids you unquestionably will have to refill are coolant and engine oil.

Because the oil galleries of a newly-assembled LS are dry of oil, all engines should ideally be "prelubed" using special equipment. Unfortunately, the design of the LS oil pump does not allow an easy way to do this. In the old days of the Gen I small-block and its distributor-driven pump, prelubing was a simple matter of inserting a drill-operated tool into the distributor hole. But because the pumps of Gen III/IV engines drive off of the crank snout, the oil pump can't turn separately from the rest of the engine. You should not be overly concerned about this issue, though; most of the time, the assembly lubricants used during final assembly should adequately protect the friction surfaces in the engine during the initial seconds before oil pressure builds.

The extra-cautious reader may consider priming the engine's oil galleries by attaching a pump to this port in the front driver side of the engine. This can be as simple as a hand-operated pump, so long as you have a fitting with the correct threads. (Depending on your underhood configuration, this may be easier to do before installing the engine into your vehicle.) Once you do this, make sure to reinstall the oil gallery plug with fresh Teflon-based sealant on its threads.

This brings us to an important topic that we've alluded to earlier, but never had the chance to fully explain: the *type* of oil to use during initial startup is critical. It is recommended that you use the thickest weight oil you can find that will still work with the bearing clearances you've set on your rods and mains. For most engines, the same SAE 30 you used to lubricate parts during final assembly will provide the best protection during startup and the initial miles of operation before your first oil change. Also, this must be a conventional oil—do *not* use synthetic oils until you have accumulated at least a few thousand miles on your new engine; more on that momentarily. Also, if you have used all the right assembly lubricant in all the right places inside your engine, no fancy oil additives will be needed at startup (although as laws and regulations force oil manufacturers to alter their formulations to include less anti-wear additives, this may become necessary one day soon!).

Start-Up

It's about that time! The moments just before your new motor is started for the first time are easily the most nerve-racking in any engine build. Rest assured, though: if you have taken the time to carefully pre-assemble your engine, been meticulous during final assembly, and double-checked everything under the hood after installation, chances are that you won't have any problems at startup. Take one last look that oil, coolant, and other fluid levels are topped off. Turn on the ignition, but do not engage the starter; look and listen for anything suspicious. Do this a few times to cycle the fuel pump and build pressure in the fuel rails (they're probably dry, having been disconnected for so long). This will allow you to check for any fuel leaks. In the exercise of complete caution, you should have a fire extinguisher and battery disconnect wrench close at hand—you can never be too safe!

With the above final items checked, the moment of truth has arrived. Turn the starter; the engine should begin to run almost immediately. Watch that oil pressure comes up within the first few seconds, and do not operate the throttle pedal—your EFI system should command the engine to idle for you. (For high-performance rebuilds that will require extensive custom tuning, you must at least install a "rough" computer calibration that will enable the engine to run for the first time—we'll discuss more about tuning in a moment.) As the engine starts to warm up, listen for any possible mechanical issues inside the engine (for example, lifters that have failed to pump up), and watch for fuel or other fluid leaks that could arise without warning, at any time. Sight, sound, and smell are all important here, and if you notice anything odd, shut the engine down immediately and figure out what's wrong.

When the coolant reaches full temperature (indicating that the thermostat has opened), turn the engine off and let it cool awhile. You will then want to top off the coolant; some vehicles require special procedures for this, so refer to your service manual if you are unsure. This may need to be done a few times, as air bubbles may take a few run cycles to find their way out of the system. You should also inspect the oil at this time to make sure it's still at the full level and to ensure no foreign substances have found their way into it (this could indicate a number of problems, so call your machine shop if you see anything—they're a good source of advice for pretty much any startup-related issue you might experience).

Break-In

During the first miles of operation, the parts inside your new engine will mate to each other during the process we all know as break-in. Despite the fact that today's engine parts are manufactured to more exacting tolerances than ever before, there is still a microscopic uniqueness to any one piece of metal that comes off of an assembly line or out of a machinist's vertical hone. Think of this mating like the last step in machining: all

The "CKP Learn" Procedure

There is one ECM/PCM-related concern that can arise with any engine rebuild, be it stock or high-performance. While not true in all situations, sometimes a so-called Crankshaft Position System Variation Learn Procedure (a.k.a. "CKP Learn" or "CASE Learn") must be performed once your rebuilt engine has been reinstalled into the vehicle. Basically, this procedure "calibrates" your ECM or PCM to particular component characteristics that influence your engine's CKP sensor, including the likes of the reluctor ring, crankshaft, harmonic balancer, and even the sensor itself. Tiny physical variations in each of these items are what the procedure is designed to compensate for. Because no two engines are exactly the same in this department, the procedure is also often required when swapping ECMs or PCMs between vehicles.

We're mentioning this procedure now because sometimes, the CKP system variation information stored in your ECM or PCM (which was calibrated to your old engine) is so out-of-whack compared to the "new" variations in your rebuilt LS, that the engine may barely run at all. Other times, the information is good enough that the engine will run, but it may perform poorly and trigger a DTC.

The good news is that many tuning tools on the market can perform the CKP Learn procedure, and the same goes for all good tuning shops. (We will be discussing both tuning tools and shops momentarily.) In a worst-case scenario, you can even take your vehicle to a GM dealer to have them perform the procedure for you. It's fairly simple and will only take a few minutes of your time. Don't let a failure to understand the possible need for this procedure hold up the success of your rebuild project!

surfaces that contact one another finish each other to the point they'll operate at for the useful life of the engine. The problem is that if we try to rush this initial wear-in, engine parts will instead wear out prematurely.

Even though advents like the LS's rollerized valvetrain mean break-in procedures are far less involved than small-block engines of decades past, the following tips should be followed to ensure long component life:

Use a variety of engine speeds

Frequently vary your engine RPM and load for at least the first 500 miles, and avoid high-RPM usage and full-throttle operation during this time. Conversely, don't "baby" the engine constantly at very low-throttle, low-RPM conditions.

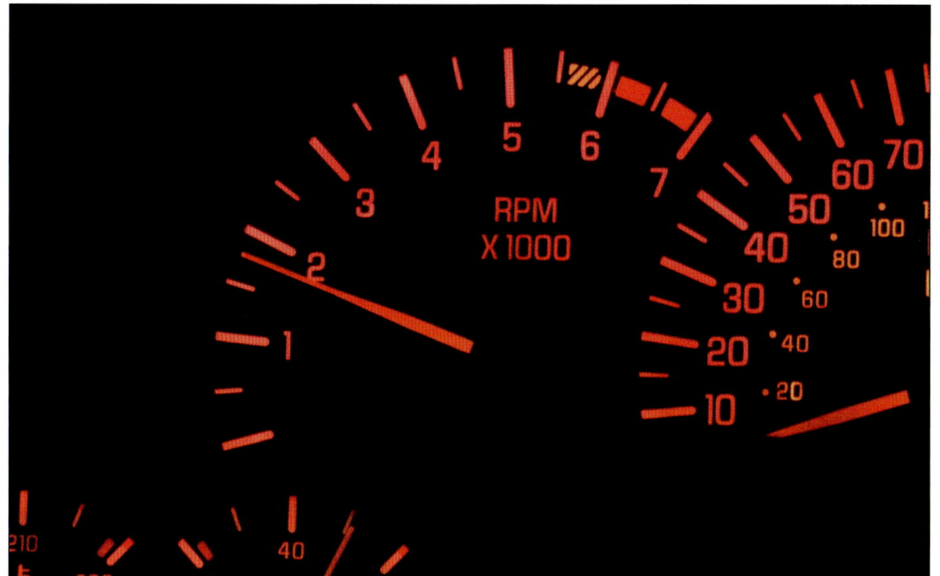

A close eye kept on your tachometer is in order during break-in. You certainly don't want to send the RPM zinging to redline constantly, but you also don't want to "lug" the engine at very low speeds. Either of these scenarios would create stresses on your internal engine parts that you should not subject them to yet.

Keep the oil clean

A substantial amount of metal particulate (from rings, bearings, and other surfaces) will find its way into the engine oil during break-in, and this is normal. Keep the engine oil clean by using a high-quality oil filter and by changing the oil after 100, 500, and 1,500 miles. You can then switch to a normal oil change schedule.

No synthetics

Do not use synthetic oil for these first few oil changes. It is important that piston rings wear in the correct manner to form a proper seal with the cylinder walls; if this does not happen, it will greatly compromise efficiency and horsepower. If synthetics are used too soon, they may never allow full ring seal to occur; and in severe cases, they can even require tearing an engine down to deglaze the cylinders! Use of synthetics *after* break-in is not only perfectly permissible, but strongly recommended for long engine life and increased engine efficiency. (Please note that some LS engines received a "factory fill" of synthetic oil when new. GM and other OEMs specifically tailor cylinder wall finishes and piston ring types to allow use of synthetics from day one; because these characteristics in your rebuilt engine are most likely different, you should not do the same.)

ECM/PCM Tuning

The days of simple carburetor jet swaps and distributor twists resulting in a properly-tuned small-block are long gone. Today, the same parameters are controlled by a complex set of sensors and computerized control unit(s) that make up an electronic fuel injection system. While it takes a little more know-how to properly adjust such a system, there is little argument that modern EFI systems make high-output engines much more driver-friendly, fuel efficient, and—perhaps best of all—produce more torque and horsepower. We simply do not have the space here to discuss the ins-and-outs of tuning these systems, but here's a

Tuning Shops

A chassis dynamometer is by far the best way to safely evaluate adjustments to your vehicle computer's calibration, and any good tuning shop will have one, along with data acquisition equipment inclusive of (at a minimum) a wideband oxygen sensor system. Dynos are not the full tuning picture, though, as most high-performance rebuilds will require adjustments to correct drivability problems like a wavering idle or tip-in hesitation–common issues that can only be properly evaluated out on the road.

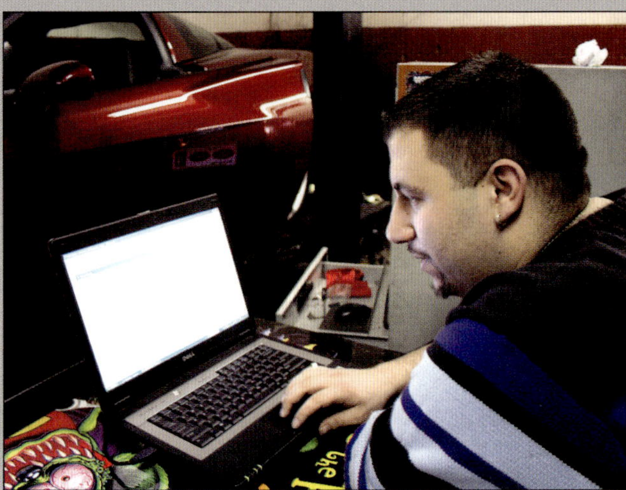

Believe it or not, the full-throttle corrections performed on a dyno are relatively easy in comparison to adjusting the myriad factors that affect part-throttle engine operation. In general, a much greater time commitment and more highly skilled tuner is needed to achieve parameters like closed-loop AFR and fuel injector duty-cycle targets. Here, Matt Sorian of TT Performance Parts has just completed a full-throttle dyno tune of an LS2 engine and makes preliminary adjustments using his tuning suite of choice in preparation for a computer reflash and road test.

There are many shops throughout the country that cater to high-performance GM cars and trucks. You probably have at least one or two of them near your hometown; they're mainly in the business of installing bolt-on parts and head/cam packages, and some even rebuild LS engines. Obviously, anyone who has read through this book has taken care of these issues on their own, but high-performance shops also offer something even the most avid do-it-yourselfer will appreciate: tuning.

These establishments have specialists on staff who are trained in tuning the on-board computers of LS-equipped vehicles, and that's what we're interested in here. The exact tuning needs of your rebuilt LS engine will vary, but well-equipped shops have the capability to not only make part-throttle adjustments to correct for driveability problems on the street, but put the engine completely through its paces safely on a chassis dynamometer. Vehicles equipped with automatic transmissions often require modifications to the transmission calibration as well, and these shops can adjust parameters like shift firmness, shift RPM, and other characteristics like downshift behavior.

Several tiers of tuning packages are generally available at most tuning shops to match the level of modifications your LS engine has undergone, and while these services aren't free, they are affordable and worth their weight in gold.

quick rundown of the basic issues you may need to deal with.

While stock rebuilds will require no changes to ECM or PCM calibration (the possibility of a CKP Learn excepted), high-performance rebuilds are a different story entirely. LS engines that incorporate the likes of extra cubic inches, higher-flow heads, and/or larger fuel injectors will not run properly, or at all, without changes to the engine computer's internal maps. Exactly how much recalibration you will need is a question of degree. Engines that incorporate minor internal modifications like small camshafts (or external bolt-ons like exhaust headers) may not require many changes, but some recalibration will still need to be performed. In this case, you may consider picking up a so-called tuning suite such as those made by EFILive or HP Tuners.

While the core features of most tuning tools are fairly easy to learn and will enable you to achieve satisfactory results (if you're careful), highly-modified engines would require learning the tuning program in its entirety and constructing all engine maps from the ground up. The time commitment needed to do this would be immense, and the process would be fraught with dangers—one wrong keystroke could destroy your new engine! Probably the best option, then, for most high-performance rebuilds is to get a specialized shop to perform a full custom tune on your vehicle. See our "Tuning Shops" Workbench Tip for more information. Extremely radical engines can even eclipse the capabilities of stock engine computers, in which you may need to have such a shop install and tune a standalone aftermarket EFI system. Note, however, that the need for this type of system isn't as common as it used to be on GM high-performance vehicles—LS engine controllers are much more intelligent and adjustable than earlier iterations from the 1980s and early 1990s, and so can simply be recalibrated to match the vast majority of high-performance scenarios.

Conclusion

If you've gone through this Workbench book step-by-step, you will have a fully rebuilt (and properly tuned) LS engine at this point. But there is one final step to take, and that is to get out there and enjoy it! I sincerely hope you've found the process of rebuilding your Gen III or IV LS engine to be as rewarding as its end product. Thanks for reading. —Chris Werner

Back in Chapter 2, we hinted at the usefulness of scan tools in helping read trouble codes that could arise with your newly rebuilt engine. While dedicated scan-only tools are available, so-called handheld tuners (like this one from Hypertech) are a good option because they also provide pre-set fuel and ignition maps that are designed to improve on the factory tune. Some can also accommodate or make adjustments for certain mild modifications like external bolt-ons. For these reasons, stock or near-stock rebuilds stand to benefit significantly from one of these units.

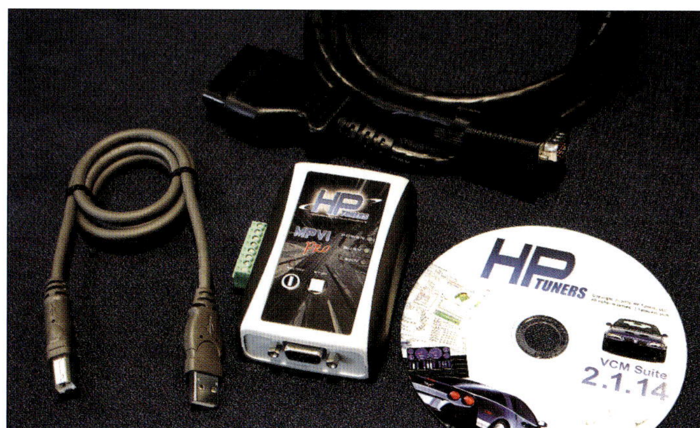

Complete tuning tools (often known as tuning suites) enable infinite adjustability of all parameters inside your vehicle's engine computer. In addition to doubling as scan tools in the traditional sense (i.e., giving the ability to read DTCs), units like this one from HP Tuners also allow you to record the outputs of all of your engine's sensors for detailed evaluation. Be cautioned that while such a tool will give you the power to make an endless array of alterations to your vehicle's computer, specialized expertise is needed to safely incorporate changes beyond minor tweaks of spark and fuel maps.

If you're going to try and dial-in the calibration of your vehicle's computer to match your newly-rebuilt high-performance engine, a wideband oxygen (O_2) sensor system is a very important thing to have. Such a system gives you the power to monitor the engine's air-fuel ratio in real time, and best of all, most can easily be wired into your tuning tool and data logged concurrently with the outputs of your engine's factory sensors. This particular air/fuel meter is made by FAST and can even be mounted on your vehicle's dashboard for easy viewing.

General Calculations for Rotating Assembly

Bore area = Pi x (Bore/2)^2

Swept volume = Bore area x Stroke
= [Pi x (Bore/2)^2] x Stroke

Engine displacement = 8 x Swept volume

Height of piston face = [Stroke/2]
+ Connecting rod length
+ Piston compression height

Deck clearance = [Deck height]
– [Height of piston face]

Notes:

-Stock deck height (the distance from the centerline of the crankshaft main bearing bore to the block deck surface) for LS engines is 9.240 +/- 0.005 inches.

-For most LS engines, deck clearance (or more appropriately, piston-to-deck clearance) will be a negative number, meaning the face of the piston protrudes from the bore slightly. It can also be measured during preassembly using a deck bridge (see "Cam Degreeing: Verification vs. Alteration" in Chapter 7 for more information).

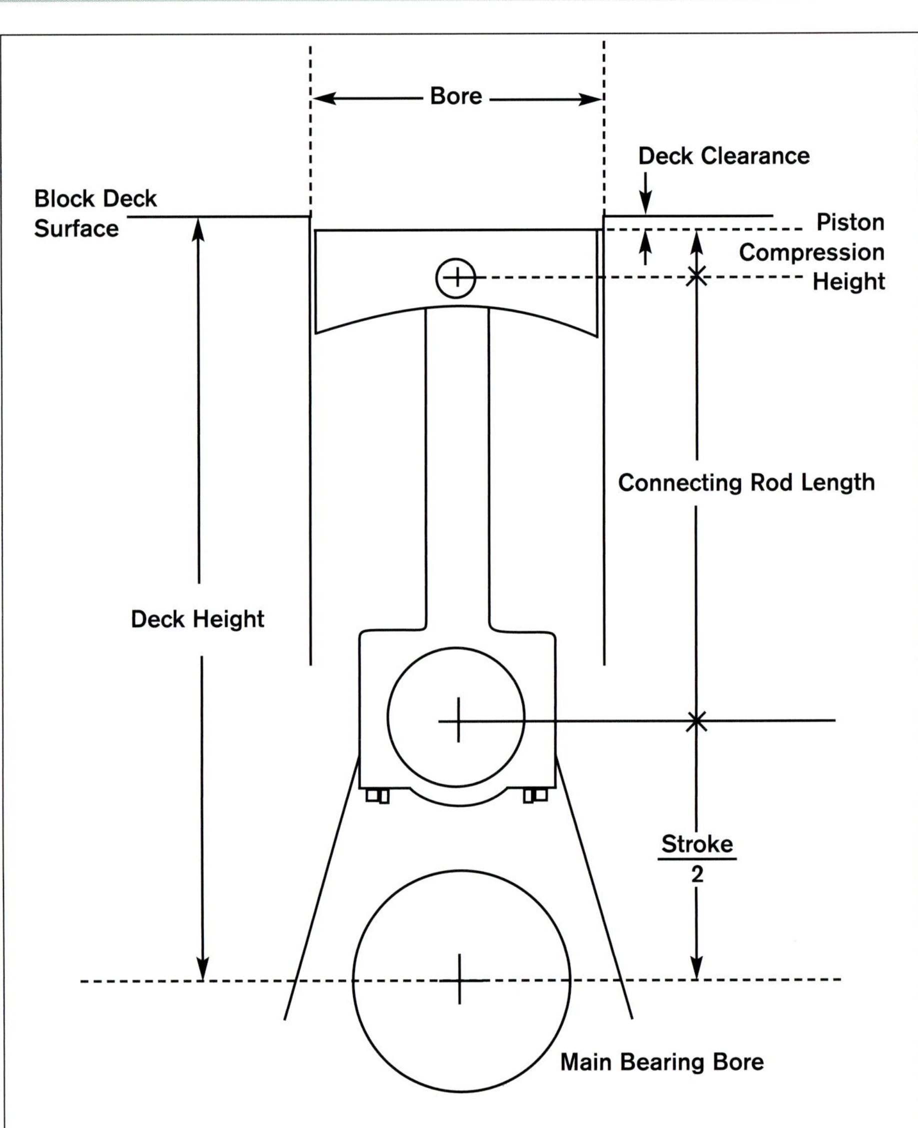

Idealized cross-sectional representation of a piston, connecting rod, and crankshaft, with the piston located at TDC.

Compression Ratio Calculation

Aside from swept volume there are several volumes that must be known for calculation of compression ratio. As shown in Figure 2, they are as follows:

Combustion chamber volume (yellow)

Supplied by cylinder head manufacturer or can be determined by your machine shop.

Head gasket volume (blue)

Equal to the compressed thickness of the head gasket multiplied by the circular area cut out in the head gasket. Typically, this area is of slightly larger diameter than the engine's cylinder bore. Consult your head gasket manufacturer for these two values.

Deck clearance volume (orange)

Equal to the deck clearance (see the previous page) multiplied by the bore area.

Piston valve relief pocket volume (purple)

Supplied by piston manufacturer or can be determined by your machine shop. For domed or dished pistons, include volume of dome or dish here (the value will be negative for domed pistons).

Crevice volume (green)

This is the volume above the top compression ring. Difficult to measure accurately, it is often assumed to be 1 cc.

Total chamber volume =
Combustion chamber volume
+ Head gasket volume
+ Deck clearance volume
+ Piston valve relief pocket volume
+ Crevice volume

$$\text{Compression Ratio} = \frac{(\text{Total chamber volume}) + (\text{Swept volume})}{\text{Total chamber volume}}$$

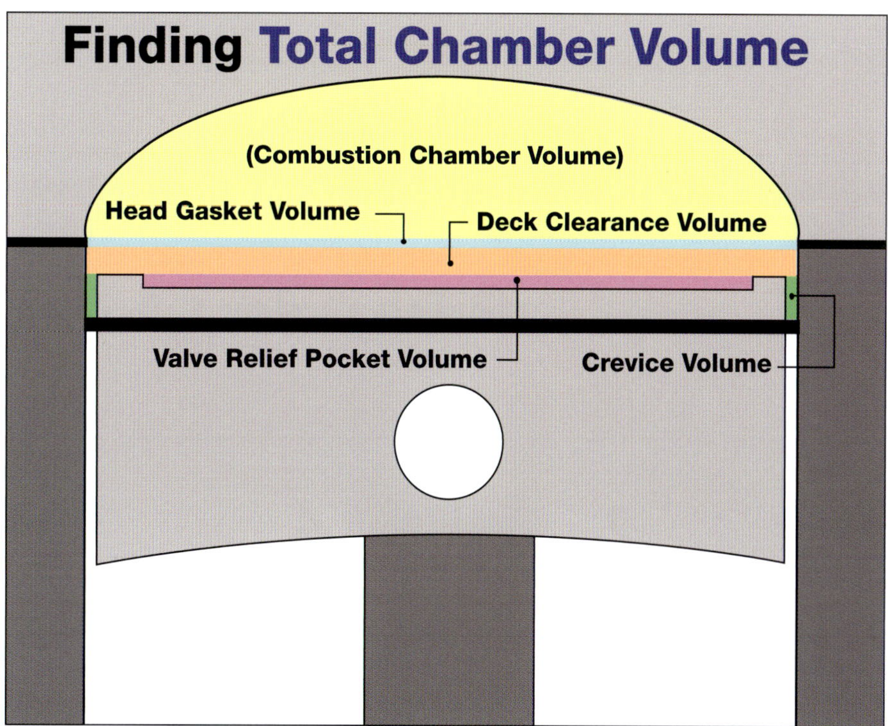

Figure 2: Idealized cross-sectional representation of a piston at TDC and the accompanying volumes that affect compression ratio.

Suggested Compression Ratios for Various Applications

Application	Compression Ratio
Forced Induction, Supercharger or Turbocharger	7.5:1 – 10.0:1
Naturally Aspirated, Stock or Mild High-Performance	9.5:1 – 11.0:1
Naturally Aspirated, Moderate to Extreme High-Performance	10.5:1 – 12.5:1

Note: the above compression ratios are general suggestions only. Consult your machine shop and/or tuning shop for more detailed guidelines.

Cylinder Numbers and Firing Order

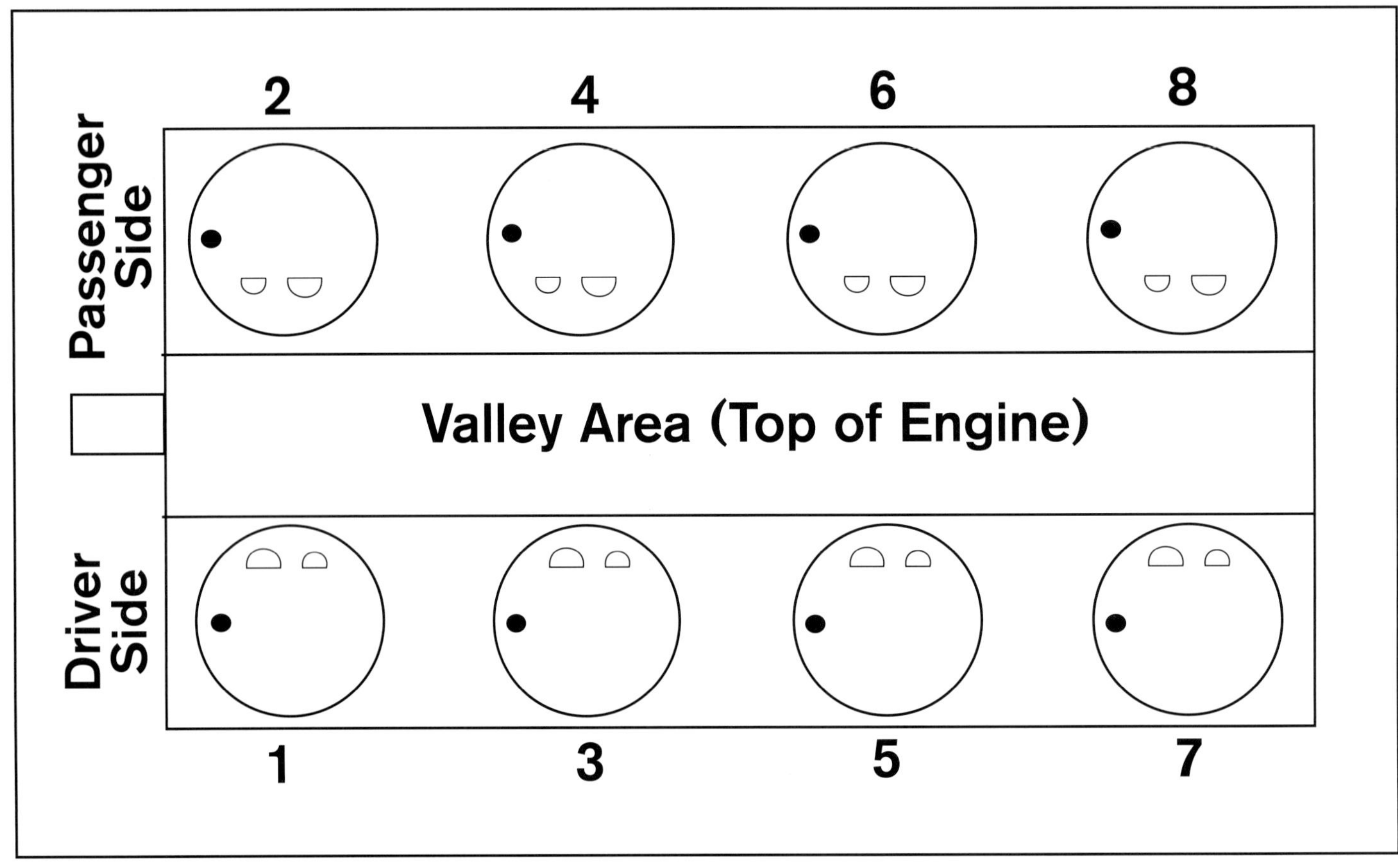

This diagram shows the locations of cylinders 1 through 8 and their firing order.

Torque Sequences

Factory main cap numbering:

Tighten inner main bearing cap bolts in this order:

Tighten outer main bearing cap bolts in this order:

Tighten head bolts in this order:

Tighten intake mainifold in this order:

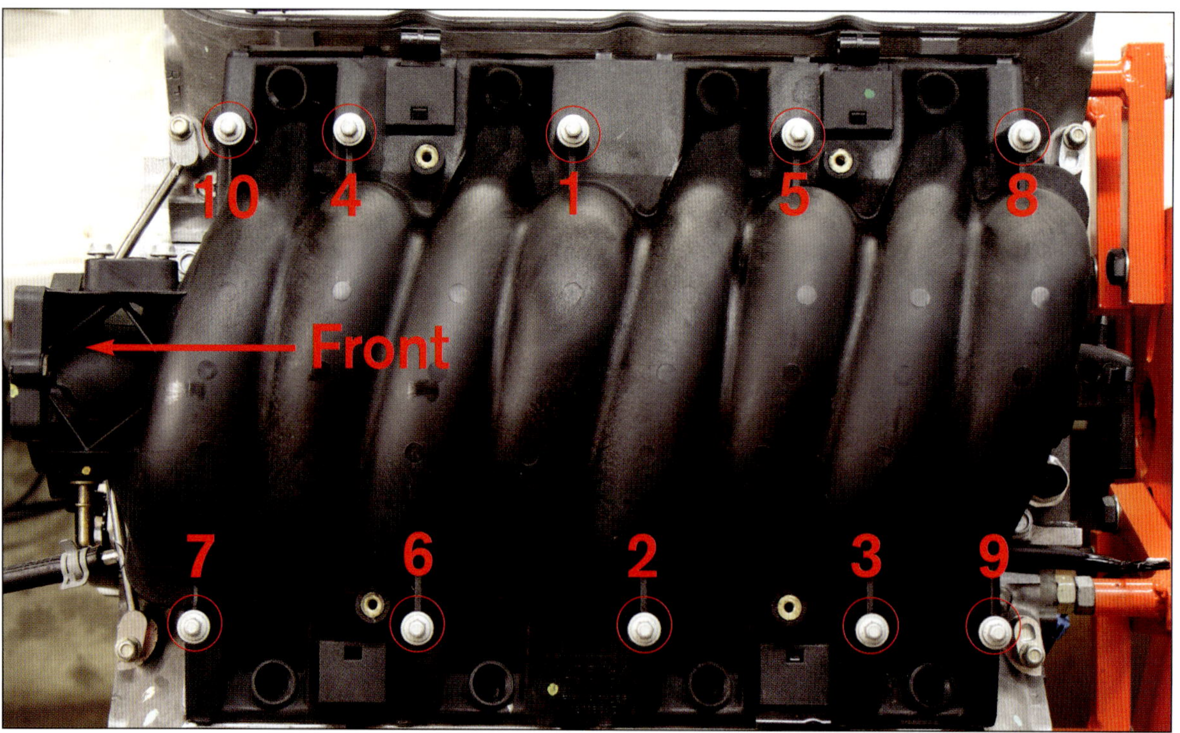

Torque Specifications

These specifications are provided for reference only. While they are correct for most engines, consult your GM service manual for the exact specifications for your LS. Always use the tightening sequences specified in Chapter 8 and elsewhere in this book. Also, remember to always use the tightening instructions provided by the manufacturer when using aftermarket parts/fasteners.

Cam retainer plate bolts:
- Hex-head bolts .... 18 ft-lbs
- TORX-head bolts .... 11 ft-lbs

Cam sprocket bolts:
- Early-style, 3 small bolts .... 18-26 ft-lbs
- Later-style, 1 large bolt:
  - Initial pass .... 66 ft-lbs
  - Final pass .... 40 degrees
- Actuator solenoid valve (VVT-equipped engines):
  - Initial pass .... 48 ft-lbs
  - Final pass .... 90 degrees

Cam sensor retainer bolt (Gen III) .... 18 ft-lbs
Cam sensor retainer bolt (Gen IV) .... 106 in-lbs

Connecting rod bolts:
- Initial pass .... 15 ft-lbs
- Final pass:
  - Early-design bolts .... 60 degrees
  - Later-design bolts .... 75 degrees

Coolant air bleed pipe/cover bolts .... 106 in-lbs
Coolant temperature sensor .... 15 ft-lbs

Crank bolt:
- Installation verification pass using OLD bolt .... 240 ft-lbs
- Initial pass using NEW bolt .... 37 ft-lbs
- Final pass .... 140 degrees

Crankshaft oil deflector nuts .... 18 ft-lbs
Crankshaft position sensor retainer bolt .... 18 ft-lbs

Cylinder head bolts (M11):
- Initial pass .... 22 ft-lbs
- Second pass .... 90 degrees
- Final pass:
  - Pre-2004 blocks; long bolts .... 90 degrees
  - Pre-2004 blocks; short bolts (2 per head) .... 50 degrees
  - 2004 and later blocks (all M11 bolts) .... 70 degrees

Cylinder head bolts (M8) .... 22 ft-lbs
Cylinder head coolant plug .... 15 ft-lbs
Dipstick tube retainer bolt .... 18 ft-lb
Engine block coolant plugs .... 44 ft-lbs
Engine block heater .... 30 ft-lbs
Engine block oil gallery plugs .... 44 ft-lbs

Engine cover bolts:
- Front cover .... 18 ft-lbs
- Rear cover .... 18 ft-lbs
- Valley cover or LOMA .... 18 ft-lbs

Exhaust manifold bolts:
- Initial pass .... 11 ft-lbs
- Final pass .... 15-18 ft-lbs

Flywheel bolts (6-bolt crank flange only):
- Initial pass .... 15 ft-lbs
- Second pass .... 37 ft-lbs
- Final pass .... 74 ft-lbs

Fuel rail bolts .... 89 in-lbs

Ignition coils:
- Coil-to-bracket bolts .... 89-106 in-lbs
- Bracket-to-valve-cover bolts .... 106 in-lbs

Intake manifold bolts (N/A supercharger):
- Initial pass .... 44 in-lbs
- Final pass .... 89 in-lbs

Knock sensor:
- Gen III .... 15 ft-lbs
- Gen IV retainer bolt .... 15-18 ft-lbs

Lifter guide tray retainer bolts .... 89-106 in-lbs

Main cap bolts (M10)
- Inner bolts – initial pass .... 15 ft-lbs
- Inner bolts – final pass .... 80 degrees
- Outer bolts – initial pass .... 15 ft-lbs
- Outer bolts – final pass .... 51-53 degrees

Main cap side bolts .... 18 ft-lbs
Oil filter fitting .... 40 ft-lbs

Oil pan:
- Drain plug(s) .... 18 ft-lbs
- M8 bolts to block/front cover .... 18 ft-lbs
- M6 bolts to rear cover .... 106 in-lbs

Oil pressure sensor:
- Gen III (threads into block) .... 15 ft-lbs
- Gen IV (threads into valley cover or LOMA) .... 26 ft-lbs

Oil pump mounting bolts .... 18 ft-lbs

Oil pump pickup tube (wet sump engines only):
- Nut holding to main cap .... 18 ft-lbs
- Bolt holding to oil pump .... 106 in-lbs

Rocker arm bolts .... 22 ft-lbs

Spark plugs:
- New spark plugs w/crush washer .... 15 ft-lbs
- Previously installed spark plugs w/crush washer .... 11 ft-lbs
- Taper-seat spark plugs .... 11 ft-lbs

Thermostat housing bolts .... 11 ft-lbs
Throttle body bolts .... 89-106 in-lbs
Timing chain dampener bolts .... 18 ft-lbs
Valve cover bolts .... 106 in-lbs

Water pump bolts:
- Initial pass .... 11 ft-lbs
- Final pass .... 22 ft-lbs

Piston Ring Gap Alignment

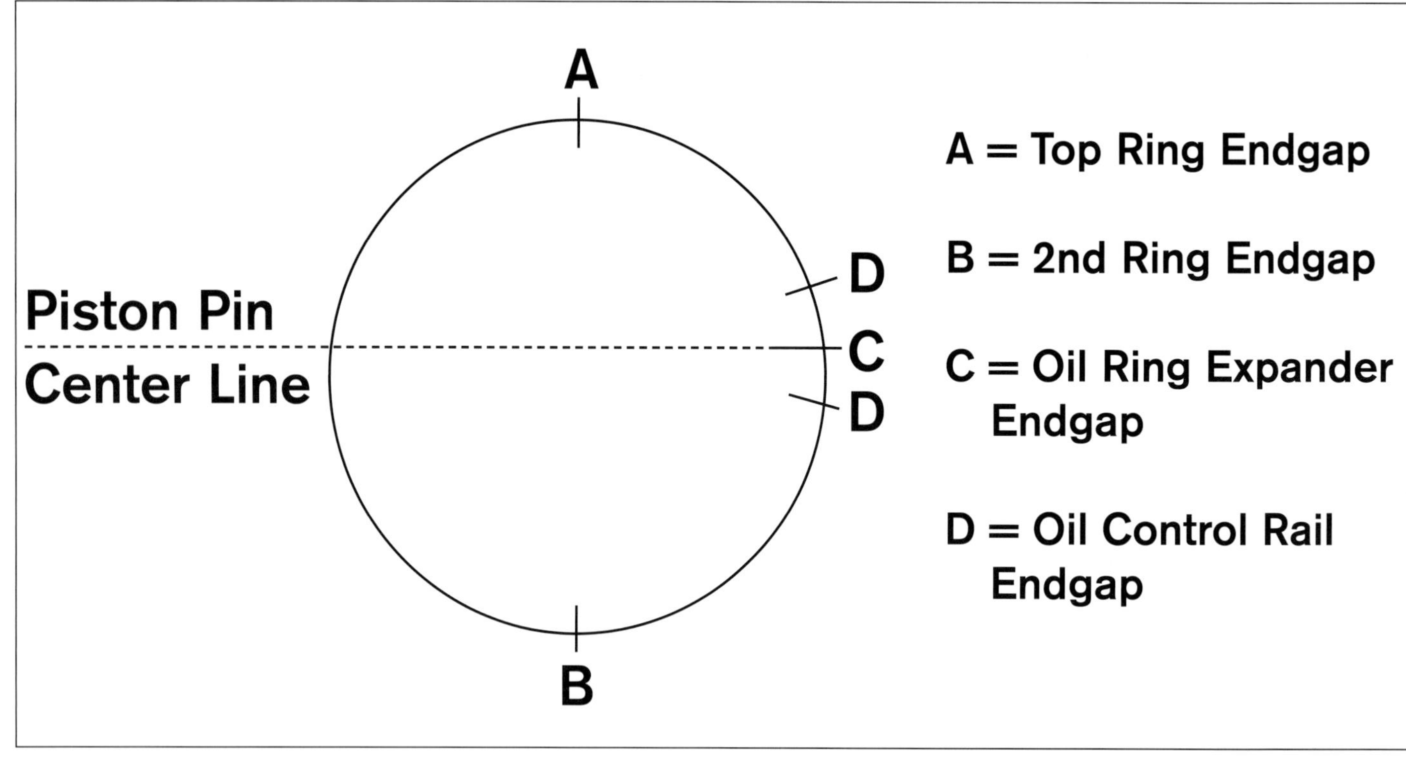

Follow this diagram during the steps of piston ring installation. See page 114 for more details.

Timing Belt/Chain Alignment Marks

Timing Chain Alignment Marks

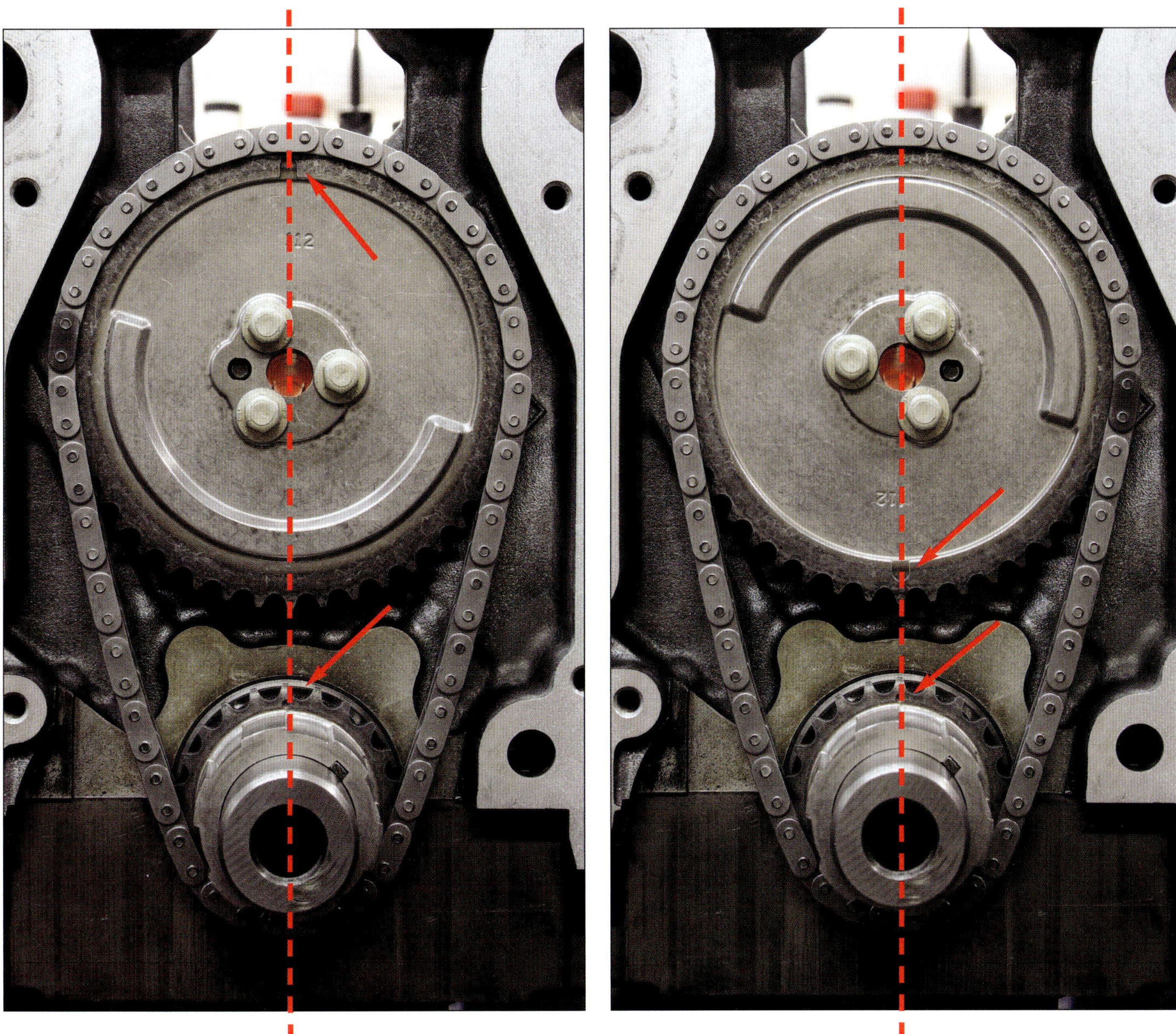

On the left, the arrows show that both the crank and cam sprocket markings are in the 12:00 position. On the right, the arrows show that the crank sprocket marking is at 12:00 but the cam sprocket marking is at 6:00. See page 100 for more details.

General Engine Specifications

Common Name	Model Years	RPO	Displacement, Liters (CI)	Bore, mm (in)	Stroke, mm (in)	Block Material	Head Material	Compres-sion Ratio	Horsepo wer**	Notes
GEN III:										
Vortec 4.8L	1999-2007	LR4	4.8 (293.4)	96.01 (3.780)	83.00 (3.268)	FE	AL	9.5:1	285	
Vortec 5.3L	1999-2007	LM7	5.3 (325.2)	96.01 (3.780)	92.00 (3.622)	FE	AL	9.5:1	295	
	2002-2007	L59	5.3 (325.2)	96.01 (3.780)	92.00 (3.622)	FE	AL	9.5:1	295	Flex Fuel
	2003-2004	LM4	5.3 (325.2)	96.01 (3.780)	92.00 (3.622)	AL	AL	9.5:1	300	
	2005-2007	L33	5.3 (325.2)	96.01 (3.780)	92.00 (3.622)	AL	AL	9.9:1	310	
LS1	1997-2004	LS1	5.7 (345.7)	99.00 (3.898)	92.00 (3.622)	AL	AL	10.1:1	350	The original Gen III
LS6	2001-2005	LS6	5.7 (345.7)	99.00 (3.898)	92.00 (3.622)	AL	AL	10.5:1	405	Higher-output version of the LS1
Vortec 6.0L	1999-2007	LQ4	6.0 (364.1)	101.6 (4.000)	92.00 (3.622)	FE	FE/AL	9.4:1	335	Pre-2001 versions had iron heads
	2002-2007	LQ9	6.0 (364.1)	101.6 (4.000)	92.00 (3.622)	FE	AL	10.0:1	345	
GEN IV:										
Vortec 4.8L	2007-2009	LY2	4.8 (293.4)	96.01 (3.780)	83.00 (3.268)	FE	AL	9.1:1	295	
	2010-	L20	4.8 (293.4)	96.01 (3.780)	83.00 (3.268)	FE	AL	8.8:1	302	Flex Fuel
Vortec 5.3L	2005-2009	LH6	5.3 (325.2)	96.01 (3.780)	92.00 (3.622)	AL	AL	9.9:1	315	AFM (most versions)
	2007-2009	LY5	5.3 (325.2)	96.01 (3.780)	92.00 (3.622)	FE	AL	9.9:1	320	AFM
	2007-	LMG	5.3 (325.2)	96.01 (3.780)	92.00 (3.622)	FE	AL	9.9:1	326	AFM; Flex Fuel
	2007-	LC9	5.3 (325.2)	96.01 (3.780)	92.00 (3.622)	AL	AL	9.9:1	326	AFM; Flex Fuel; VVT (some versions)
	2008-2009	LH8	5.3 (325.2)	96.01 (3.780)	92.00 (3.622)	AL	AL	9.9:1	300	
	2008-	LMF	5.3 (325.2)	96.01 (3.780)	92.00 (3.622)	FE	AL	9.9:1	310	Flex Fuel; VVT (some versions)
	2010-	LH9	5.3 (325.2)	96.01 (3.780)	92.00 (3.622)	AL	AL	9.7:1	300	VVT
LS4	2005-2009	LS4	5.3 (325.2)	96.01 (3.780)	92.00 (3.622)	AL	AL	10.0:1	303	AFM; used in FWD cars
LS2	2005-2009	LS2	6.0 (364.1)	101.6 (4.000)	92.00 (3.622)	AL	AL	10.9:1	400	
Vortec 6.0L	2007-	LY6	6.0 (364.1)	101.6 (4.000)	92.00 (3.622)	FE	AL	9.6:1	360	VVT
	2007-2009	L76	6.0 (364.1)	101.6 (4.000)	92.00 (3.622)	AL	AL	9.6:1*	367	AFM; VVT (truck versions)
	2008-2009	LFA	6.0 (364.1)	101.6 (4.000)	92.00 (3.622)	AL	AL	10.8:1	332	Hybrid, AFM, VVT, LIVC
	2010-	L96	6.0 (364.1)	101.6 (4.000)	92.00 (3.622)	FE	AL	9.6:1	360	Flex Fuel (some versions); VVT
	2010-	LZ1	6.0 (364.1)	101.6 (4.000)	92.00 (3.622)	AL	AL	10.8:1	332	Hybrid, AFM, VVT, LIVC
Vortec 6.2L	2007-2008	L92	6.2 (376.0)	103.25 (4.065)	92.00 (3.622)	AL	AL	10.5:1	403	VVT
	2009-	L9H	6.2 (376.0)	103.25 (4.065)	92.00 (3.622)	AL	AL	10.4:1	403	Flex Fuel; VVT
	2010-	L94	6.2 (376.0)	103.25 (4.065)	92.00 (3.622)	AL	AL	10.4:1	403	AFM; Flex Fuel; VVT
LS3	2008-	LS3	6.2 (376.0)	103.25 (4.065)	92.00 (3.622)	AL	AL	10.7:1	436	
LSA	2009-	LSA	6.2 (376.0)	103.25 (4.065)	92.00 (3.622)	AL	AL	9.0:1	556	Supercharged
LS9	2009-	LS9	6.2 (376.0)	103.25 (4.065)	92.00 (3.622)	AL	AL	9.1:1	638	Supercharged; dry sump oiling system
L99	2010-	L99	6.2 (376.0)	103.25 (4.065)	92.00 (3.622)	AL	AL	10.4:1	400	AFM; VVT
LS7	2006-	LS7	7.0 (427.6)	104.78 (4.125)	101.6 (4.000)	AL	AL	11.0:1	505	Dry sump oiling system

**10.4:1 on car versions.*
***Estimate; varies by vehicle and model year. Where appropriate, the highest horsepower rating the engine received as of MY 2010 is listed*
Note: The above designations pertain to the United States market only, but the same or very similar engines are used in other North American markets as well as overseas.

Helpful Unit Conversions

Constants: Pi = 3.14159

Distance: 1 inch = 25.4 mm

Displacement (Volume): 1 ci (cubic inch) = 16.3871 cc
1 liter = 1,000 cc

Torque: 1 ft-lb. = 1.3558 N-m
1 ft-lb. = 12 in-lb.

Glossary of Common Abbreviations

ABDC - After Bottom Dead Center. *See* BDC.

AFM - Active Fuel Management (GM terminology). Same as DOD.

AFR - Air-Fuel Ratio.

ATDC - After Top Dead Center. *See* TDC.

BDC - Bottom Dead Center. Refers to a piston having reached its minimum height in the bore.

BBDC - Before Bottom Dead Center. *See* BDC.

BTDC - Before Top Dead Center. *See* TDC.

CMM - Coordinate Measuring Machine.

CNC - Computerized Numeric Control. An advanced type of machining utilizing computer-controlled and operated tooling. Also often used to refer to the machine itself.

DOD - Displacement On Demand (GM terminology). Same as AFM.

DOHC - Double Overhead Cam.

DTC - Diagnostic Trouble Code.

ECM - Engine Control Module. Essentially identical to ECU.

ECU - Engine Control Unit. Essentially identical to ECM.

EFI - Electronic Fuel Injection.

EGR - Exhaust Gas Recirculation. An emissions control device used on some vehicles.

EGT - Exhaust Gas Temperature.

FWD - Front-Wheel Drive.

IAC - Idle Air Control. A small electric motor commanded by the engine computer that adds or subtracts airflow to adjust idle speed.

IAFM - Integrated Air-Fuel Module (GM terminology). Comprises the intake manifold, throttle body, fuel rails, and fuel injectors.

IAT - Intake Air Temperature. Also used to refer to the Intake Air Temperature sensor.

ICL - Intake Centerline. A measure used to describe camshafts.

I.D. - Inside Diameter.

LIVC - Late Intake Valve Closure. Feature incorporated into certain Gen IV engines.

LOMA - Lifter Oil Manifold Assembly (GM terminology). Present on engines equipped with AFM.

LSA - Lobe Separation Angle. A measure used to describe camshafts. (Also, RPO desisnation of a Gen IV Variant.)

M8, M11, etc. - Indicates the size of a metric bolt's threaded portion (8mm, 11mm, etc.) This is NOT the same as the size of wrench needed to turn the bolt's head.

MAF - Mass Air Flow sensor.

MAP - Manifold Absolute Pressure. Also used to refer to the Manifold Absolute Pressure sensor.

MIL - Malfunction Indicator Lamp.

NPT - National Pipe Tapered thread.

OBD-II - On-Board Diagnostics. The II designation refers to the revised program of OBD that was mandated beginning in 1996.

O.D. - Outside Diameter.

OE - Original Equipment.

OEM - Original Equipment Manufacturer.

OHC - Overhead Cam.

OHV - Overhead Valve. Refers to engines using pushrod valve actuation.

PCM - Powertrain Control Module. Incorporates both ECM and TCM into a single unit on some vehicles.

PCV - Positive Crankcase Ventilation. An emissions control device.

PE - Power Enrichment. A common parameter used in engine computers.

RA - Roughness Average.

RPO - Regular Production Order.

RWD - Rear-Wheel Drive.

SAE - Society of Automotive Engineers.

SBC - Small-Block Chevrolet.

TCM - Transmission Control Module.

TCS - Traction Control System. Can alter spark advance, throttle position, and other engine-related parameters to control wheelspin.

TDC - Top Dead Center. Refers to a piston having reached its maximum height in the bore.

TPI - Tuned Port Injection. The GM port fuel injection system introduced in the 1980s to which the EFI systems of all LS engines owe their roots.

TPS - Throttle Position Sensor. Also often used to refer to the voltage output of this sensor.

VCM - Vehicle Control Module.

VVT - Variable Valve Timing (GM terminology).

WOT - Wide Open Throttle.

SOURCE GUIDE

Automotive Racing Products (ARP)
1863 Eastman Avenue
Ventura, CA 93003
(800) 826-3045
www.arp-bolts.com

Bill Ceralli Competition Engines
395 East 18th Street
Paterson, NJ 07524
(973) 742-4972

Cometic Gasket
8090 Auburn Road
Concord, OH 44077
(800) 752-9850
www.cometic.com

COMP Cams
(a division of COMP Performance Group)
3406 Democrat Road
Memphis, TN 38118
(901) 795-2400
www.compcams.com

Fuel Air Spark Technology (FAST)
(a division of COMP Performance Group)
3400 Democrat Road, Suite 110
Memphis, TN 38118
(901) 260-FAST
www.fuelairspark.com

Federal-Mogul
(Fel-Pro, Sealed Power, Speed-Pro, etc.)
www.federal-mogul.com

Goodson Tools and Supplies for Engine Builders
156 Galewski Drive, P.O. Box 847
Winona, MN 55987-0847
(800) 533-8010
www.goodson.com

GM Performance Parts
www.gmperformanceparts.com

Helm
14310 Hamilton Ave.
Highland Park, MI 48203
(800) 782-4356
www.helminc.com

Hypertech
3215 Appling Road
Bartlett, TN 38133
(901) 382-8888
www.hypertech.com

Jesel
1985 Cedar Bridge Ave
Lakewood, NJ 08701
(732) 901-1800
www.jesel.com

LRB Performance Machine Company
22-B Lasinski Road
Franklin, NJ 07416
(973) 209-7770
www.lrbperformance.com

Lunati, LLC
11126 Willow Ridge Drive
Olive Branch, MS 38654
(662) 892-1500
www.lunatipower.com

Patriot Performance
103 Rainbow Industrial Blvd.
Rainbow City, AL 35906
(888) 462-8276
www.patriot-performance.com

Racing Head Service (RHS)
(a division of COMP Performance Group)
3416 Democrat Road
Memphis, TN 38118
(877) 776-4323
www.racingheadservice.com

School of Automotive Machinists
1911 Antoine
Houston, TX 77055
(713) 683-3817
www.samracing.com

SLP Performance Parts
1501 Industrial Way North
Toms River, NJ 08755
(732) 349-2109
www.slponline.com

Helpful LS-oriented websites:
www.LS1Tech.com
www.GMHighTechPerformance.com